METHODS IN MOLECULAR BIOLOGY

For further volumes:
http://www.springer.com/series/7651

Crop Breeding

Methods and Protocols

Edited by

Delphine Fleury and Ryan Whitford

Australian Centre for Plant Functional Genomics (ACPFG), University of Adelaide, Urrbrae, SA, Australia

Humana Press

Editors
Delphine Fleury
Australian Centre for Plant Functional Genomics (ACPFG)
University of Adelaide
Urrbrae, SA, Australia

Ryan Whitford
Australian Centre for Plant Functional Genomics (ACPFG)
University of Adelaide
Urrbrae, SA, Australia

Additional material to this book can be downloaded from http://extras.springer.com

ISSN 1064-3745 ISSN 1940-6029 (electronic)
ISBN 978-1-4939-0445-7 ISBN 978-1-4939-0446-4 (eBook)
DOI 10.1007/978-1-4939-0446-4
Springer New York Heidelberg Dordrecht London

Library of Congress Control Number: 2014936846

Printed on acid-free paper

Humana Press is a brand of Springer
Springer is part of Springer Science+Business Media (www.springer.com)

Preface

One of our greatest challenges is to feed several billion people in a changing environment whereby water and nutrients are predicted to become limiting yet the demands of population growth continue. The first green revolution enabled us to increase crop production through the breeding of new hybrid and semi-dwarf varieties, the use of fertilizers and modern agronomical practices. The recent advent of molecular technologies has changed the way plant breeders identify and select their germplasm, genetic variation can now be assessed at the DNA level, with much of this information finding application to their selection strategies. A small size breeding program can easily utilise 20–50 markers for their molecular-assisted selection, focusing on known traits such as flowering time, grain quality or disease resistance. Large breeding programs are now routinely assessing millions of molecular data points every year in order to identify new genes and diagnostic markers. This information facilitates genetic background selection of progenies, therefore allowing the best choice of plants for the following round of selection. Furthermore, the molecular identity of gene sequences underlying important traits can now be used in the creation of novel transgenic varieties.

This volume addresses breeders and pre-breeding researchers in the crop science community. The first two chapters give guidelines on how to design a breeding strategy for the selection of an ideal variety or genetic ideotype, and how to transform gene sequence information into practical diagnostic markers. The second section (Chapters 3–7) provides protocols for breeders using molecular markers in selection programs and for laboratories providing molecular services to breeding programs. The methodologies collated were selected based on cost, efficiency and applicability to both medium- and large-scale breeding systems. These protocols can therefore suit different needs and capacity: from small laboratories analysing molecular markers on a one-by-one basis to the increasingly popular high-throughput protocols for high-capacity laboratories. Molecular biology and breeding now involve considerable analysis in silico, from data collection, storage to complex statistical analysis. The third section (Chapters 8–12) describes statistical programs and software to aid implementation of molecular data into breeding programs. The fourth and final section (Chapters 13–19) describes methodologies that facilitate the generation of genetic diversity and its characterisation, for example creating new alleles by introgression, mutagenesis and plant transformation, as well as capturing and fixing this variation through doubled haploidy.

We would like to thank the authors for their kind contribution. The described methods are those that they have either developed or used within their own breeding programs. The detailed guidelines and tutorials were developed so that the methods could be easily adopted by breeders. We hope this volume will help in expanding the use of molecular technologies for the creation of tomorrow's crop varieties.

Urrbrae, SA, Australia

Delphine Fleury
Ryan Whitford

Contents

Contributors

GREGORIO ALVARADO • *Biometrics and Statistics Unit of the Crop Research Informatics Laboratory, International Maize and Wheat Improvement Center (CIMMYT), Mexico DF Mexico*

JEFFREY ANTHONY • *LGC Genomics Ltd, Hoddesdon, Herts, UK*

NICOLAS BARDOL • *INRA, UMR de Génétique Végétale, Ferme du Moulon, Gif sur Yvette, France; Euralis Semences, Domaine de Sandreau, Mondonville, France*

GUYLAINE BLANC • *INRA, UMR de Génétique Végétale, Ferme du Moulon, Gif sur Yvette, France*

JULIEN BONNEAU • *Australian Centre for Plant Functional Genomics, School of Botany, The University of Melbourne, Melbourne, VIC, Australia*

SUE BROUGHTON • *Department of Agriculture and Food WA, South Perth, WA, Australia*

CORINNE CALLEGARI • *Australian Centre for Plant Functional Genomics, University of Adelaide, Glen Osmond, SA, Australia; School of Agriculture, Food and Wine, Waite Research Institute, University of Adelaide, Hartley Grove, Urrbrae, SA, Australia*

JESÚS J. CERÓN-ROJAS • *Biometrics and Statistics Unit of the Crop Research Informatics Laboratory, International Maize and Wheat Improvement Center (CIMMYT), Mexico DF Mexico*

SCOTT CHAPMAN • *CSIRO Plant Industry, St. Lucia, QLD, Australia*

ALAIN CHARCOSSET • *INRA, UMR de Génétique Végétale, Ferme du Moulon, Gif sur Yvette, France*

JOSÉ CROSSA • *Biometrics and Statistics Unit of the Crop Research Informatics Laboratory, International Maize and Wheat Improvement Center (CIMMYT), Mexico DF, Mexico*

PHILIP A. DAVIES • *South Australian Research and Development Institute, Adelaide, SA, Australia*

MARK DIETERS • *School of Agriculture & Food Sciences, The University of Queensland, St Lucia, QLD, Australia*

CHONGMEI DONG • *Plant Breeding Institute, University of Sydney, Narellan, NSW, Australia*

SERIK ELIBY • *Australian Centre for Plant Functional Genomics, University of Adelaide, Glen Osmond, SA, Australia; School of Agriculture, Food and Wine, Waite Research Institute, University of Adelaide, Hartley Grove, Urrbrae, SA, Australia*

STEPHEN J. FLETCHER • *Chemistry and Molecular Biosciences, The School of Biological Sciences, University of Queensland, Brisbane, QLD, Australia*

DELPHINE FLEURY • *Australian Centre for Plant Functional Genomics (ACPFG), University of Adelaide, Urrbrae, SA, Australia*

MELISSA GARCIA • *Australian Centre for Plant Functional Genomics, The University of Adelaide, Glen Osmond, SA, Australia*

FRANCK GAUTHIER • *INRA, UMR de Génétique Végétale, Ferme du Moulon, Gif sur Yvette, France*

JOHN C. HARRIS • *Australian Centre for Plant Functional Genomics, University of Adelaide, Glen Osmond, SA, Australia; School of Agriculture, Food and Wine, Waite Research Institute, University of Adelaide, Hartley Grove, Urrbrae, SA, Australia*

ADRIAN HATHORN • *CSIRO Plant Industry, St. Lucia, QLD, Australia*

MATTHEW HAYDEN • *Department of Primary Industries Victoria, Victorian AgriBioscience Centre, La Trobe Research and Development Park, Bundoora, VIC, Australia*
CHUNLIN HE • *Generation Challenge Programme, c/o CIMMYT, Texcoco, Mexico*
JOHN HOLME • *LGC Genomics Ltd, Hoddesdon, Herts, UK*
BAO LAM HUYNH • *Department of Nematology, University of California, Riverside, CA, USA*
GULNUR ISKAKOVA • *Australian Centre for Plant Functional Genomics, University of Adelaide, Glen Osmond, SA, Australia; School of Agriculture, Food and Wine, Waite Research Institute, University of Adelaide, Hartley Grove, Urrbrae, SA, Australia*
AINUR ISMAGUL • *Australian Centre for Plant Functional Genomics, University of Adelaide, Glen Osmond, SA, Australia; School of Agriculture, Food and Wine, Waite Research Institute, University of Adelaide, Hartley Grove, Urrbrae, SA, Australia*
JEAN-LUC JANNINK • *Department of Plant Breeding and Genetics, USDA-ARS, Robert W. Holley Center for Agriculture and Health, Cornell University, Ithaca, NY, USA*
JOHANN JOETS • *INRA, UMR de Génétique Végétale, Ferme du Moulon, Gif sur Yvette, France*
KELVIN KHOO • *School of Agriculture, Food and Wine, Waite Research Institute, University of Adelaide, Glen Osmond, SA, Australia*
NATALIYA KOVALCHUK • *Australian Centre for Plant Functional Genomics, School of Agriculture, Food and Wine, Waite Research Institute, University of Adelaide, Hartley Grove, Urrbrae, SA, Australia*
DIANE E. MATHER • *School of Agriculture, Food and Wine, Waite Research Institute, University of Adelaide, Glen Osmond, SA, Australia*
IRYNA MAZONKA • *Australian Centre for Plant Functional Genomics, University of Adelaide, Glen Osmond, SA, Australia; School of Agriculture, Food and Wine, Waite Research Institute, University of Adelaide , Hartley Grove, Urrbrae, SA, Australia*
GRAHAM MOORE • *Crop Genetics Department, John Innes Centre, Norwich Research Park, Norwich, UK*
LAURENCE MOREAU • *INRA, UMR de Génétique Végétale, Ferme du Moulon, Gif sur Yvette, France*
MARK A. NEWELL • *The Samuel Roberts Noble Foundation, Ardmore, OK, USA*
SERGÍO PEREZ-ELIZALDE • *Biometrics and Statistics Unit of the Crop Research Informatics Laboratory, International Maize and Wheat Improvement Center (CIMMYT), Mexico DF, Mexico*
LE PHUOC THANH • *Australian Centre for Plant Functional Genomics, The University of Adelaide, Glen Osmond, SA, Australia; Department of Plant Protection, College of Agriculture and Applied Biology, Can Tho University, Can Tho city, Vietnam*
PETER SHARP • *Plant Breeding Institute, University of Sydney, Narellan, NSW, Australia*
PARMINDER K. SIDHU • *South Australian Research and Development Institute, Adelaide, SA, Australia*
RICHARD M. TRETHOWAN • *Plant Breeding Institute, The University of Sydney, Sydney, NSW, Australia*
ELISE J. TUCKER • *Australian Centre for Plant Functional Genomics Waite Research Institute, School of Agriculture, Food and Wine, University of Adelaide, Glen Osmond, SA, Australia*
FABIO VALENTE • *INRA, UMR de Génétique Végétale, Ferme du Moulon, Gif sur Yvette, France*
JOHN WOODWARD • *DuPont Pioneer, Johnston, IA, USA*

Chapter 1

Defining a Genetic Ideotype for Crop Improvement

Richard M. Trethowan

Abstract

While plant breeders traditionally base selection on phenotype, the development of genetic ideotypes can help focus the selection process. This chapter provides a road map for the establishment of a refined genetic ideotype. The first step is an accurate definition of the target environment including the underlying constraints, their probability of occurrence, and impact on phenotype. Once the environmental constraints are established, the wealth of information on plant physiological responses to stresses, known gene information, and knowledge of genotype × environment and gene × environment interaction help refine the target ideotype and form a basis for cross prediction.

Once a genetic ideotype is defined the challenge remains to build the ideotype in a plant breeding program. A number of strategies including marker-assisted recurrent selection and genomic selection can be used that also provide valuable information for the optimization of genetic ideotype. However, the informatics required to underpin the realization of the genetic ideotype then becomes crucial. The reduced cost of genotyping and the need to combine pedigree, phenotypic, and genetic data in a structured way for analysis and interpretation often become the rate-limiting steps, thus reducing genetic gain. Systems for managing these data and an example of ideotype construction for a defined environment type are discussed.

Key words Ideotype, Crop improvement, Genotype × environment interaction, Gene × environment interaction, Gene effects, Cross prediction

1 Introduction

Plant breeders traditionally base selection on phenotypes generated in the target environments, and many have established ideotypes to help focus the selection process. The ideotype concept was first suggested by Donald [1]. He proposed the establishment of a hypothetical plant based on traits thought to enhance yield potential as a guide to selection. However, the process of developing a target genotype or ideotype is complex, and the purpose of this chapter to provide a road map to the establishment of a refined genetic ideotype.

The first step in the process of defining an accurate genetic ideotype is a complete definition of the target environment including all the underlying stresses and constraints, their probability of

Delphine Fleury and Ryan Whitford (eds.), *Crop Breeding: Methods and Protocols*, Methods in Molecular Biology, vol. 1145, DOI 10.1007/978-1-4939-0446-4_1, © Springer Science+Business Media New York 2014

occurrence, and impact on phenotype. There is extensive information on production statistics in most regions, soil type, historical temperature and rainfall patterns, and market requirements. These data can be used to identify most probable environment types and to prioritize traits for breeding and selection [2]. Once the environmental constraints are established the wealth of information on plant physiological responses to stresses, genetic variation, and known gene information for disease resistance, phenology, plant architecture, and grain quality can be used to develop a target ideotype for the target region.

Most plant breeding programs have extensive historical multi-environment trial data available, and the response of genotypes across the target environment and the extent of genotype × environment interaction can be assessed. This change in genotype rank across the target area can help redefine the target environment or refine the broader environment into sub-environments based on genotype responses. While these analyses show us the extent of genotype × environment interaction and environmental data may help explain some of the variance, there remains a significant portion of total genotype × environment interaction that is unexplained.

As we understand more about the genetic control of key traits and more quantitative trait loci (QTLs) and genes are discovered with linked or diagnostic markers, it is possible to estimate gene or QTL × environment interactions, thus improving our understanding of genotype × environment interaction and further refining our definition of ideotype. These gene effects can also be used to estimate epistasis through cumulative analysis, thus providing the basis for cross prediction in wheat breeding [3]. Nevertheless, much of the variation in key traits such as grain yield remains unexplained once known gene effects have been estimated. Association genetic analysis based on random genotyping can help "backfill" regions of the chromosome where no known genes of major effect are located [4]. These chromosomal regions can then be recombined in crosses further defining the genetic ideotype.

Once the genetic ideotype is defined using the above tools the challenge remains to build the ideotype in a plant breeding program. There are a number of strategies that can be employed, and some of these, such as marker-assisted recurrent selection (MARS) and genomic selection, also provide valuable information for the optimization of genetic ideotype. However, as our understanding of genetic ideotype improves the informatics required to underpin the realization of this target becomes crucial. The reduced cost of genotyping and the need to combine pedigree, phenotypic, and genetic data in a structured way for analysis and interpretation often become the rate-limiting steps, thus reducing genetic gain.

This chapter outlines the steps required for the practical determination of genetic ideotype using wheat as an example. In the process strategies for assembling and refining the target genotype are discussed.

2 Understanding the Target Environment

2.1 The Physical Production Environment: The Concept of Mega-Environments

Most plant breeders work within a defined region or set of environment types. Superficially, the plant breeder can define the production environment in terms of soil type, latitude, altitude, rainfall, disease expression, cropping patterns, management practices, and market requirements, all of which help define the underlying stresses and the gene combinations needed to influence genotype adaptation. The wheat breeding program at the International Maize and Wheat Improvement Center (CIMMYT) targets wheat production areas in developing countries globally, and a series of mega-environments have been defined based on these parameters to better target breeding and selection and the subsequent deployment of germplasm [5, 6]. Using these definitions 12 global mega-environments were identified, 6 representing spring wheat and 6 winter wheat-growing areas; these were further divided on the basis of rainfall, irrigation, growing season temperature patterns, soil acidity, and latitude. The CIMMYT program then developed broad ideotypes for each environment type based on knowledge of the underlying stresses. For example, germplasm targeting the high-rainfall environments (mega-environment 2) would have high levels of rust, leaf blight, and head blight resistance; would be red seeded and thus tolerant to preharvest sprouting; and would be lodging resistant.

2.2 Redefining the Concept of Mega-Environment

Recently, this mega-environment definition has been fine-tuned using geographic information systems (GIS) [2]. Spatial data, such as the average temperature of the coolest quarter of the year were used to redefine wheat-growing environments in India. Previously, genetic materials were targeted to optimal irrigated areas or areas affected by heat stress based on earlier definitions of mega-environments [6]. However, once the quarterly temperature was mapped and overlaid with production statistics and trial site locations it became obvious that some trial locations were misclassified. Thus the targeting of germplasm for evaluation in India could be refined. These authors also used spatial data to predict future cropping scenarios based on climate change. The area subject to heat stress was predicted to increase significantly on the Indian subcontinent. These predictions will influence the "future" genetic ideotype for these regions, and the plant breeder must increase the focus on heat-adaptive traits, such as temperature stress tolerance per se, early flowering, and improved spot-blotch resistance if yields are to be maintained or improved over time. GIS also helps predict the movement of pathogens such as rust disease in wheat [7]. In this instance the movement of a new stem rust strain, Ug99, was predicted based on cropping patterns, survey data, and prevailing winds. Thus the future genetic constitution of cultivars in the infection pathway can be informed by this early warning tool.

Remote sensing images from satellites can be used to assess and confirm land use and land cover patterns [8] and even soil type distribution [9]. These images in combination with other spatial data will help fill gaps in our knowledge of key wheat production regions.

2.3 Definition of a Target Population of Environments Based on Modelling

Proper definition of the underlying constraints and the most probable season types is essential for the targeted development of cultivars. Understanding these constraints will help the plant breeder manage genotype × environment interaction and ultimately improve the heritability of selection for complex characters such as yield. This classification is often referred to as the target population of environments, which is defined as "the set of environments in which cultivars can be grown within the geographical area targeted by a breeding programme" [10]. A recent attempt to characterize Australian northeastern wheat-growing environments used the APSIM model and historical climatic and soil data to characterize the larger geographic area and local field experiments [11]. Three primary environment types based on water deficit during flowering and grain filling were identified; of these terminal drought stress accounted for 50 % of environments. These findings were then used to reduce the impact of genotype × environment interaction, thus increasing genetic variance and better predicting cultivar performance. In a variable rainfall environment such as northeastern Australia where genotype × year interactions comprise a significant component of total variance it is difficult for the plant breeder to target a particular plant ideotype. However, better understanding of the dominant environment types based on these types of simulation will allow the breeder to assign weights to individual trials and traits, identify the most responsive germplasm for crossing, and develop an approximate genetic ideotype for selection.

3 Defining a Genetic Ideotype

Physiologists have attempted to define target ideotypes for plant breeders based on physiological traits. One such attempt is presented in Fig. 1 and represents an optimized genotype for drought and heat resistance [12]. The model assumes that each of these primary traits are genetically independent and that no compensatory effects influence their expression. The authors also estimated the yield gains achievable from optimization of the physiological responses at different stages of development based on several years of field-based experimentation. Each of these primary traits can potentially be broken down into subsidiary traits under genetic control such as carbon isotope discrimination or water-soluble carbohydrates that influence water-use efficiency (WUE) or early plant growth.

However, from a practical plant breeding viewpoint it is possible to define physiological traits that influence yield into four categories:

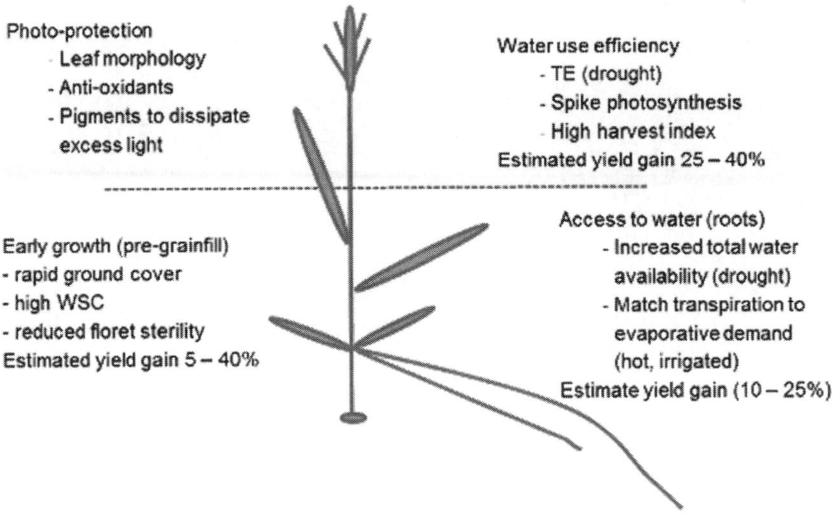

Fig. 1 A conceptual plant ideotype (adapted from 12)

those that influence emergence and establishment, early growth and development, pre-flowering characters, and post-flowering characters including those important during grain filling (Table 1). Traits such as grain size and embryo size [13], coleoptile length [14], and coleoptile width [15] are important for emergence and establishment, whereas early growth is controlled by vigor [16], water-soluble stem carbohydrates (WSC) [17], and rooting patterns [18]. Prior to flowering, traits such as canopy temperature depression (CTD) [19], transpiration efficiency (TE) [20], osmotic adjustment (OA) [21], WSC [22], and rooting patterns [23] are all influential. Post-flowering, a wide range of physiological traits can be assessed including WSC [24], WUE [25], leaf area duration (LAD) [26], and grain quality. Some have suggested that reduced tillering may improve yield and adaptation in environments subject to water stress at and post-anthesis [27]. These plant types would not build excessive early biomass, thereby reducing tiller loss and subsequent small grain size from the impacts of stress later in the growth cycle. Many of these traits are difficult to measure and can only be used to assess parents prior to crossing. Others, such as grain size, coleoptile length, vigor, CTD, tiller number, and LAD, can be easily assessed and used to select among segregating progeny.

While some of the physiological traits listed in Table 1 are difficult or time consuming to assess, the identification of QTL and linked markers would significantly improve selection efficiency, thereby making selection within segregating generations possible. Table 2 lists some recent reports of QTL linked to these key traits. A feature of most reports is the large number of QTLs of relatively minor individual effect detected. Ibrahim et al. [28] report

Table 1
Physiological traits considered to be important in controlling yield in water-limited environments

Emergence and establishment	Early growth	Pre-flowering (booting)	Post-flowering
Coleoptile length [14, 15]	Vigor [16]	CTD [19]	Grain yield [48]
Grain size [13]	WSC [17]	TE/CID [20]	WUE [25]
Embryo size [13]	Rooting patterns [18, 23]	OA [21]	CTD [19]
	Reduced tillering [27]	WSC [22]	TE/CID [20]
		Rooting patterns [23]	OA [21]
			LAD [26]
			WSC [24]
			Rooting patterns [62]
			Quality
			Grain size [13]

Note: *WSC* water-soluble stem carbohydrates, *CTD* canopy temperature depression, *TE* transpiration efficiency, *CID* carbon isotope discrimination, *OA* osmotic adjustment, *LAD* leaf area duration, *WUE* water-use efficiency

Table 2
Recent reports of QTL linked to important physiological traits that influence yield in water-limited environments

Trait	References
Coleoptile length	[29, 63]
Grain size	[38]
Early vigor	[29]
Water-soluble stem carbohydrates	[31, 32]
Canopy temperature depression	[37]
Transpiration efficiency	[33]
Osmotic adjustment	[34]
Leaf area duration	[36]
Water-use efficiency	[35]
Rooting patterns	[28]
Reduced tillering	[27]

32 QTLs for a range of root morphological traits although these tended to cluster in four major groups on chromosomes 1D, 2A, 2D, and 7D. Spielmeyer et al. [29] identified a region on chromosome 6A that influenced both coleoptile length and early vigor, and Lui et al. [30] later suggested that up to 12 QTLs influence coleoptile length and subsequent plant emergence. WSC are similarly complex [31] although some major effects, such as a large QTL on 1RS in wheat carrying the 1B/1R translocation, can be found [32]. In terms of water balance, Rebetzke et al. [33] found a number of QTLs for improved transpiration efficiency that explained up to 10 % of the trait variation; Ciuca et al. [34] identified a significant effect on chromosome 7A linked to improved osmotic adjustment, and Alexander [35] reported a number of QTLs for drought tolerance including major effects on chromosomes 4AL and 7B. The stay green trait that increases LAD under stress appears to be more simply inherited with a QTL of large effect present on 4A [36]. Although other traits, such as CTD, grain size, and tiller inhibition, are more easily assessed using non-molecular approaches, significant QTLs linked to their expression have been reported [27, 37, 38]. Clearly, any QTL must be validated in the target environment in the key germplasm of relevance to that environment. However, these QTLs and genes once validated would form an integral part of the genetic ideotype construction.

The adaptation and yield of crop species across a region are largely controlled by genes that affect phenology such as photoperiod and vernalization response. In wheat these responses are controlled by three loci each, *PpdD1, Ppd2,* and *Ppd3* and *VrnA1, VrnB1,* and *VrnD1*, for photoperiod and vernalization response, respectively [39]. A number of alleles at each locus also modify response, although the primary effects of insensitivity to day length and vernalization are dominant. There are diagnostic molecular markers available for a number of these genes, and Eagles et al. [3] used four of these, *VrnA1, VrnB1, VrnD1,* and *PpdD1*, to type a large number of genotypes grown in historical multi-environment trials in southern Australia. They identified 15 of 16 possible genotypes based on allelic variation at these four loci in the set of genotypes tested and concluded that 45 % of the observed variation in heading date could be attributed to these four loci. Similarly, semidwarf plant height and the genes controlling this character in wheat are essential for high harvest index and therefore high yield. There was a misconception that non-semidwarf (tall) habit offered an advantage under drought-stressed conditions. However, this was later dispelled by the performance of near-isogenic pairs of bread wheat and durum wheat lines based on the Green Revolution dwarfing genes *Rht1-B1b* and *Rht-D1b* [40]. These isogenic pairs were yield tested in environments ranging from severely stressed to highly productive. In most cases the semidwarf genotype was superior to the tall isoline across the wide range of environments (Fig. 2). However, there is a disadvantage associated

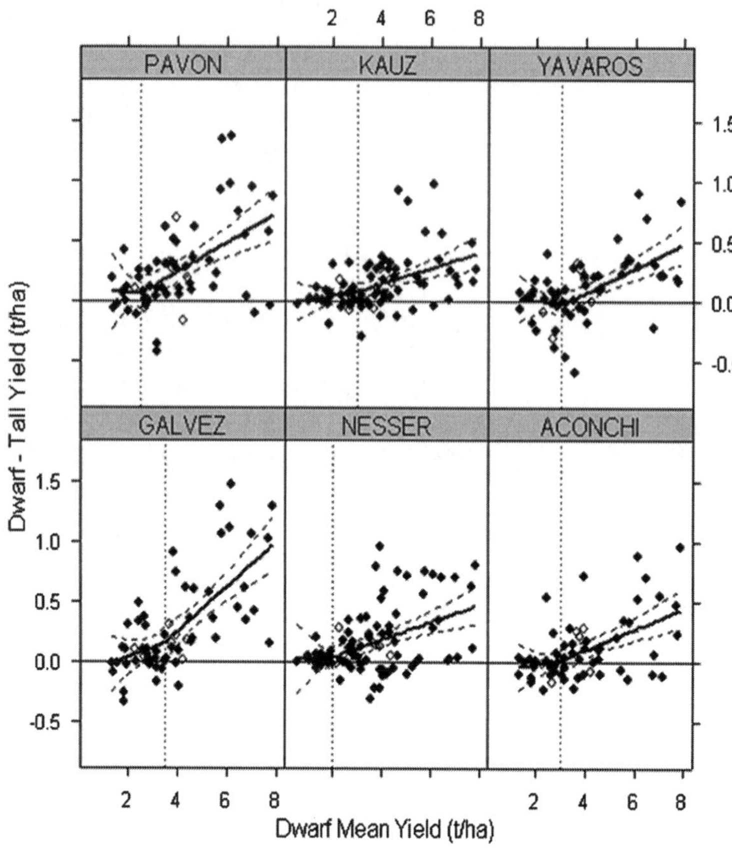

Fig. 2 The global yield performance of six isogenic pairs based on *Rht-B1b* and *Rht-D1b*. Reproduced from [40] by permission

with the *Rht-B1b* and *Rh-D1b* genes; they are gibberellic acid insensitive and therefore have shorter coleoptiles, thus limiting emergence and establishment in some environments [14]. Alternative sources of dwarfism have been discovered that are not linked to short coleoptile, and these can be introduced into breeding programs to enhance adaptation without reduction in harvest index [41]. Gibberellic acid-sensitive dwarfing genes of particular interest include *Rht 4, 5, 8, 9, 12,* and *13.* Should these dwarfing genes be de-coupled from any deleterious effects, and evidence suggests that this is possible, then the current semi-dwarfism ideotype based on *Rht-B1b* and *Rht-D1b* will change.

4 Interaction of Genotype × Environment

Once the target environment has been defined, the underlying stresses identified, and the gene combinations assembled that confer adaptation, more broadly the "must have" genes, it is necessary to both assess and exploit genotype × environment interaction.

There are many methods for analyzing genotype × environment interaction ranging from simple linear regression to more complicated techniques such as the shifted multiplicative model and pattern analysis of large unbalanced data sets. A comparison of these techniques can be found in Trethowan et al. [42]. The plant breeder aims to minimize crossover interaction, thus broadening the adaption of the materials developed and deployed.

The CIMMYT wheat program deploys genotypes developed under the mega-environment strategy in yield trials around the world each year. There have been a number of analyses of genotype × environment interaction, and these have been used to redefine mega-environments and in the process identify key locations that represent the key zones of adaptation globally [43–48]. These zones were then used to select parents and focus germplasm evaluation. According to Trethowan and Crossa [47], key locations or best predictors for yield in rainfed environments are Bethlehem in South Africa and Marcos Juarez in Argentina. These sites were identified from pattern analysis of CIMMYT's Semi-Arid Wheat Yield Trial, and squared Euclidean distances were used to find the most representative site in each primary environment cluster. Clearly, gene profiles of the materials performing well in these environments, including those controlling phenology, plant height, and disease resistance, would offer insight into the probable constitution of globally adapted wheat for water-deficit conditions. These types of analyses can also be used to identify key germplasm groupings as much as site similarities. Lillemo et al. [46] explored the performance of different gene pools within the CIMMYT germplasm targeted to high-temperature stress and found that materials with similar coefficients of parentage tended to have similar responses to high temperature.

While many analytical procedures exist to estimate the extent of interaction, it is difficult to explain why genotypes change rank from location to location within a region. The mega-environment concept mentioned earlier attempts to minimize these impacts by deploying materials targeted to specific environment types depending on the dominant stresses. However, it is often not possible to identify all key constraints within a region or an environment type based on environmental data. To overcome these limitations, probe genotypes and/or isolines that differentiate for a particular stress are effective and were recommended by Cooper and Fox [49]. More recently, Matthews et al. [50] analyzed the global performance of an adaptation trial based on lines either contrasting or near isogenic for key adaptive traits. These authors were able to use these data to better target germplasm to environments, thus significantly reducing the lines tested in key global locations. One of the isogenic contrasts in this adaptation trial is the comparison of Gatcher, which is susceptible to root lesion nematode (*Pratylenchus thornei*), with its isogenic nematode-resistant equivalent, GS50A [51]. Figure 3 shows the significant contrasts of these isolines globally. However, a positive contrast for

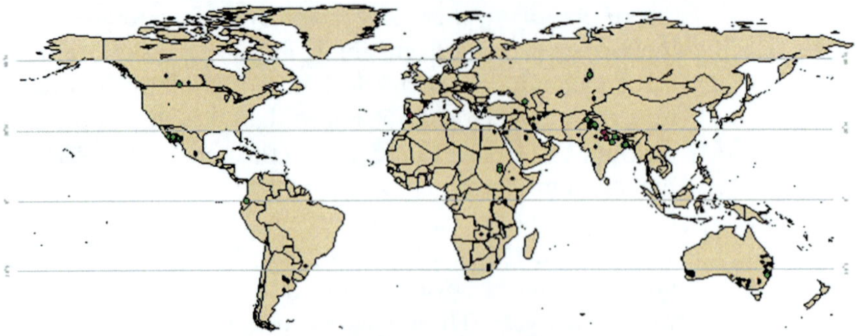

Fig. 3 The site and significance of GS50A and Gatcher isogenic pairs differing for root lesion nematode (*Pratylenchus thornei*) resistance. Reproduced from [51]

yield is suggestive and does not necessarily indicate the presence of nematodes; this would need to be confirmed with soil testing. Nevertheless, the results can be used to target soil testing to specific regions and, if the presence of nematodes is confirmed, direct the breeding and selection strategies at these locations.

The ideotype generally used by plant breeders is heavily biased to disease resistance as disease expression is easily detected, generally simply inherited, and easily linked to improved yield. In fact, much of the impact of plant breeding is linked to improved resistance in the context of pathogen mutation, and this is never more evident than in the case of rust. The rust pathogen mutates and disease resistance genes, particularly those of major or race-specific effect, break down often with disastrous consequences for farmers. However, horizontal or non-race-specific resistance as defined by Vanderplank [52] and combinations of genes will greatly reduce this impact, particularly on farmers in developing countries where the seed production systems are often ineffective and replacement cultivars are not easily developed or distributed. When defining a genetic ideotype for disease resistance the target combination of genes must be both effective and durable in the context of the farming system. In terms of genetic progress in grain yield, the rate of increase in yield over time is four times greater when disease resistance is a factor compared to advances in yield potential alone [53].

5 Interaction of Gene × Environment

As our knowledge of the inheritance of economically important traits increases so does our ability to understand gene or QTL × environment interactions. While the multiple effects of an individual gene can be estimated across large numbers of genotypes and environments, it is also possible to estimate epistatic effects by analyzing their effects in combination. Eagles et al. [3] used a mixed model analysis of photoperiod and vernalization genes characterized in 1,085 wheat genotypes grown in multi-environment trials over a 24-year period. They found that predictions of flowering were more accurate when the combinations of *Ppd* and *Vrn* genes and alleles were considered rather than summation of the individual gene effects. These authors augmented their predictions of genotype performance using environmental data to estimate vernalization saturation and using these data were able to recommend optimal sowing dates for different vernalization allelic constitutions. Clearly, it is possible to establish a target genetic ideotype using such data and optimization will depend upon our knowledge and identification of important genes. Eagles et al. [54] concluded that in southern Australia, an optimized combination of these genes is the photoperiod-insensitive allele at *Ppd-D1*, the spring allele at *Vrn-A1*, and winter alleles at the other *Vrn* loci. Interestingly many Australian cultivars with this constitution have been released by breeders without the advantage of molecular markers; the genes have been accumulated through empirical selection for yield in the target environment. Nevertheless, this information and an optimized ideotype for flowering time will help plant breeders select parents that fix these gene combinations.

An effective use of genetic effects estimated from large multi-environment data sets and subsequent gene profiling was the development of a "glutenin simulator" to predict the outcomes of crosses based on known gene effects (www.molecularplant-breeding.com). These estimations were based on characterization of glutenin and puroindoline genes and alleles across a large multi-environment data set [55]. The effects of each glutenin allele on dough strength, dough extensibility, and dough development were determined as was the effect of the puroindoline genes on grain hardness, flour water absorption, and milling yield. One of the key findings was the advantage of the puroindoline combination *Pina-D1a/Pinb-D1b* for dough extensibility, milling yield, and water absorption over *Pina-D1b/Pinb-D1b*. Interestingly, the CIMMYT wheat breeding program had a high frequency of the latter compared to Australian wheat programs reflecting the importance of milling yield in quality determinations compared to the Mexican based program. These estimations were used to develop the glutenin simulator which provides wheat breeders with a probability of obtaining the target quality phenotype (based on dough rheological properties and milling yield) from any chosen combination of glutenin and puroindoline genes.

The methods described above are essentially an association analysis using known genes and large multi-environment data sets. However, association analyses are also used to identify genomic regions linked to trait expression without prior knowledge of foreground or known genes. An overview of association analysis can be found in [56]. One such example is the association analysis conducted on global wheat data generated by the CIMMYT wheat program [4]. In this analysis significant DArT markers were detected in a number of chromosomal regions linked to yield performance and disease resistance where no genes or QTLs of major effect had previously been identified. The significant markers can be used to combine these regions by identifying complementary parents for crossing and subsequent MAS selection. Nevertheless, the estimated marker effects are only relevant if the phenotypic data is of sufficient scale to allow linkage disequilibrium and the genotypes assessed are of relevance to the breeding program using the information. For this reason association analysis of whole breeding programs is an effective strategy. The parents' intermediate and fixed line progeny can be genotyped and the phenotype generated in the multi-environment trialling systems established by most breeding programs.

6 Building the Target Genotype

Estimation of gene effects based on known genes and their epistatic interactions helps develop a genetic ideotype, and plant breeders can ensure that these gene combinations are present among their parental materials and the probability of obtaining the desired genotype estimated using predictive software. However, the inheritance of grain yield, the single most significant trait for wheat breeders, is highly complex, and known genes and QTLs are unlikely to account for a significant component of the total genetic variance. Improved and cost-effective genotyping using single-nucleotide polymorphisms (SNP) or genotyping by sequencing (GBS) now makes it possible to employ breeding strategies based on estimated marker effects to "backfill" as yet undifferentiated sections of the genome.

One such approach is MARS, and a description can be found in Charmet et al. [57]. This scheme identifies significant QTLs within each targeted cross among polymorphic parents. The progenies are generally genotyped at F3 and phenotyped at F4 or later when sufficient seed exists for multi-environment testing. Following QTL detection, the progenies are recombined on the basis of the significant marker effects (Fig. 4). Generally up to eight progeny lines can be recombined in this way using a magic [58] crossing structure. However, it is also possible to track known genes of major effect in such a scheme. The selection of parents and derived progeny could be skewed first by selection for known genes (such as photoperiod, vernalization, rust resistance, or grain quality). These genes can also be tracked during the recombination phase of a MARS approach in addition to the newly detected cross-specific QTL identified.

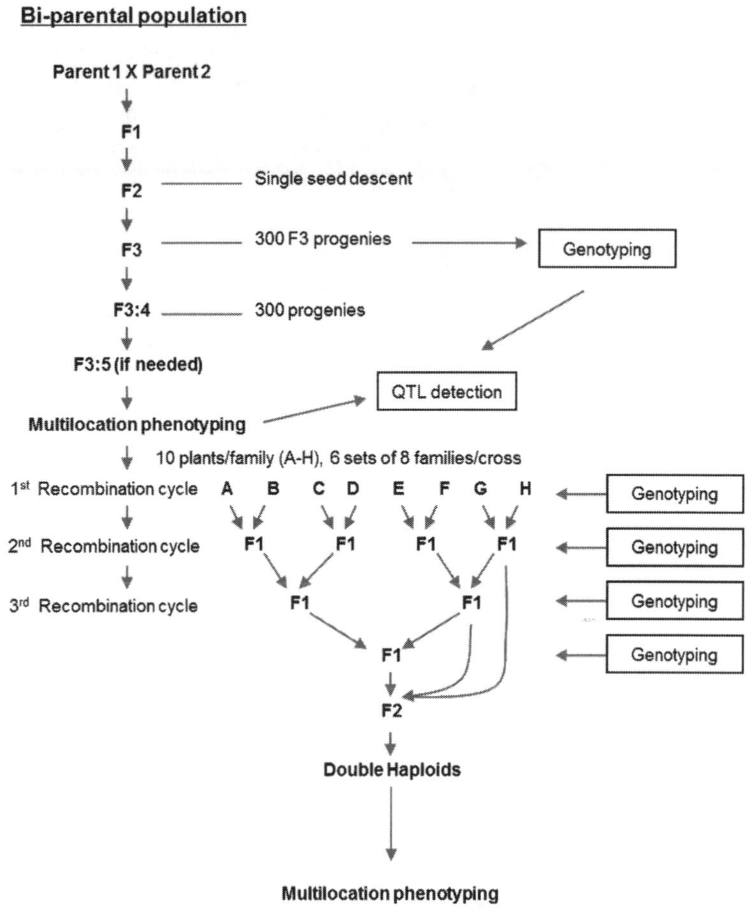

Fig. 4 A marker-assisted recurrent selection scheme. Reproduced from [64] with permission from Elsevier

Many of the current markers for known genes in wheat are SSRs. However, it should be possible to find SNP or GBS markers in these regions to further facilitate the selection process.

The major limitation of MARS is the extensive cross-specific phenotyping required to identify progeny for recombination. For this reason genomic selection (GS) offers significant advantages, and a review of GS strategies, marker systems, and techniques can be found in Paux et al. [59]. In this strategy a large training population representing germplasm important in the target population of environments is genotyped using a high-density marker system and phenotyped extensively. Following estimation of marker effects, a selection population is identified for recombination. In this strategy all the marker effects can be used regardless of significance to develop new lines with as many of the additive marker effects combined as possible. Clearly, selection for known genes of major effect such as rust resistance or grain quality can also be overlaid providing they are confirmed in the selection population.

In self-pollinated crops such as wheat the recombination phase of both MARS and GS present significant limitations as it is difficult to combine more than eight parents in a single recombination crossing strategy. However, male sterility as commonly used in hybrid seed production systems may offer some advantages. The most useful are chemical hybridizing agents (notwithstanding environmental and health safety concerns of such substances) as any genotype can be used as the female in any growing season. Of least value are the cytoplasmic male sterility systems which require the parallel development of maintainer lines, and fertility restoration genes must be present in the pollen parents. However, genetic systems such as that linked to blue seed color in wheat (60, N Darvey, personal communication) can be used as the progeny segregate for blue and white seed color. In this system the white-seeded lines are male sterile and the blue-seeded lines are fertile and therefore maintainers. To use such a system it would be necessary to convert several key backgrounds for use as females. A pollen bulk representing all potential parents could then be used to pollinate the female population, and subsequent selection for all additive markers, if on sufficient scale, should identify genotypes with optimized combinations of alleles.

Double haploids could then be made on all plants positive for the target markers to fix the materials for testing in multi-environment trials.

7 Informatics and Ideotype Construction

The key to accurately determine a genetic ideotype lies in the efficient management of the enormous amount of information encompassing pedigree, phenotypic, genetic, and environmental data available to the plant breeder. Plant breeders have gone from managing and analyzing thousands of data points to millions of data points, and the storage, access, analysis, and interpretation of these data will impact rates of genetic advance. There are various options for managing breeding program data including off-the-shelf packages such as Agrobase. Other options include locally developed systems and the International Crop Information System (ICIS) (www.icis.cgiar.org) developed through a consortium under the auspices of the International Group on International Agricultural Research (CGIAR). The ICIS program has an effective genealogy management system (GMS) that allows plant breeders to store and update pedigrees, estimate coefficients of parentage, and print and manage field books [61]. The management of pedigree data is the first step in the adoption of an integrated database. The ICIS data management system (DMS) is split into phenotypic and genotypic data storage, and all data are linked to individual pedigrees in the GMS system through

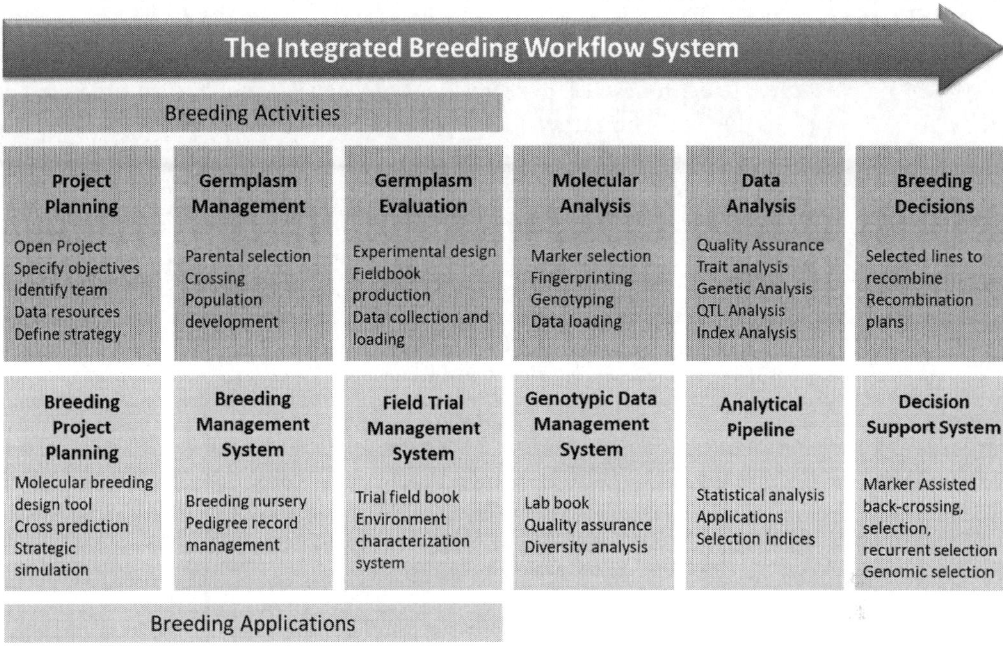

Fig. 5 Workflow in the integrated breeding platform. Reproduced from [65] with permission

unique numbers or identifiers. The DMS system is still without full functionality at the time this manuscript was written. Nevertheless, despite the number of systems available to the plant breeder, either commercially or free of charge, no one system manages the four primary data sources, pedigrees, phenotypes, genotypes, and environmental data, efficiently in a single database. An additional problem, once these data are housed, is their access, integration, and visualization, and this requires additional interrogation tools. The Integrated Breeding Platform (IBP) of the Generation Challenge Program attempts to do this using a breeding management system (www.integratedbreeding.net/integrated-breeding-%adworkbench). The IBP has developed a range of interrogation/interpretation tools to augment the flow of data through a breeding program. Systems such as these, publicly available and targeting researchers primarily in developing countries, help the plant breeder manage the complexities of molecular plant breeding from parent identification to advanced line selection and commercialization. A description of the workflow is presented in Fig. 5.

These tools and the efficient management of data cut across the various steps in genetic ideotype development. Whether it be analysis of multi-environment phenotypic data, estimating gene effects, or managing the complexities of a genomic selection program, these data can only be used effectively if stored properly and accessible in formats that allow integration.

8 Conclusion

The process of genetic ideotype development is summarized in Fig. 6. The target ideotype is not static and changes as the plant breeder gains access to better information and analysis tools. While the steps outlined in Fig. 6 and described above represent a process, an ideotype can be developed using information at any stage.

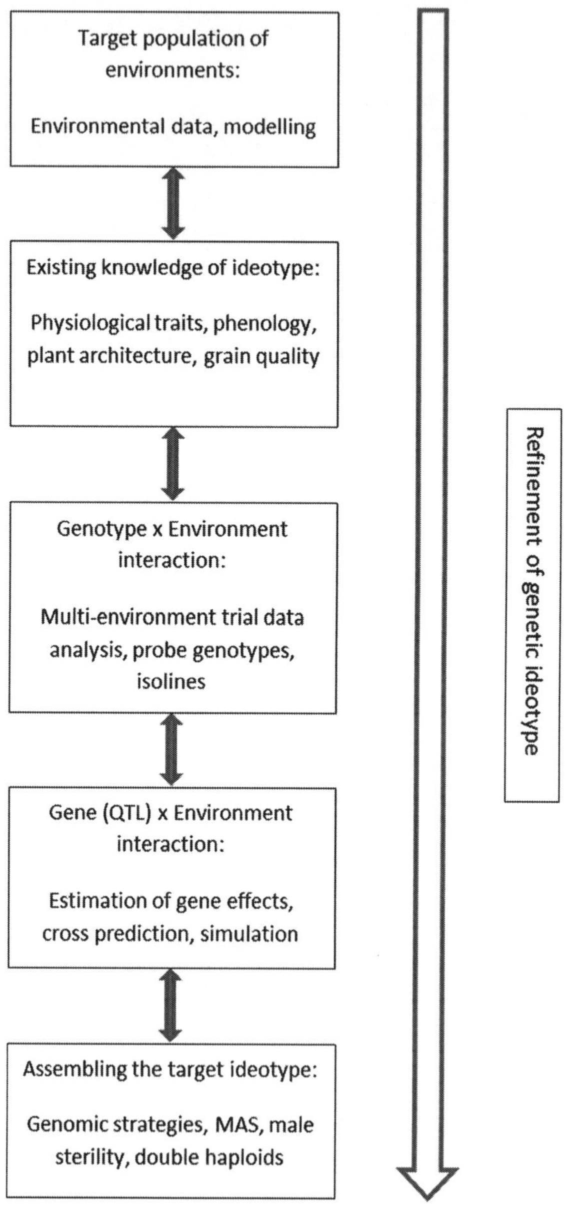

Fig. 6 Summary of the process of defining and assembling the target genetic ideotype

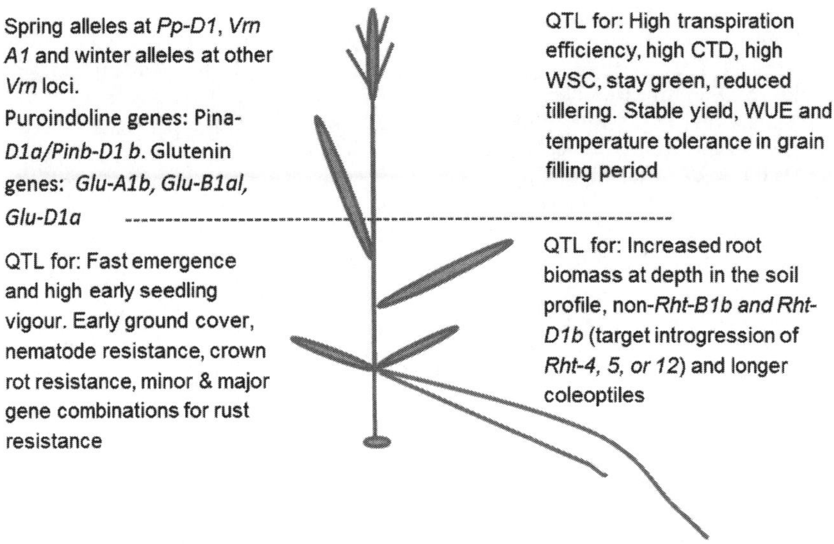

Fig. 7 An approximate trait ideotype for the northern grain-growing region of NSW and Queensland (based on currently available QTL and gene information)

Clearly, the process begins with a thorough understanding of the target environment and all underlying stresses and constraints, including markets. The definition of a proto-ideotype can then be assembled based on knowledge of traits and genes, and this is further refined by analysis of genotype and gene × environment interaction. Further refinement is then possible using genomic strategies to identify marker effects in regions of the genome where no reported genes of major effect exist.

On this basis it is possible to estimate a probable ideotype for wheat growing in a defined region, and this is attempted below for the northern winter grain-growing region of New South Wales and southern Queensland in Australia (Fig. 7). This ideotype takes us to box II of Fig. 6 (existing knowledge of the target ideotype) and is therefore superficial, based only on known trait data, and assumes yield stability across the target environment. However, it is a starting point for the accurate establishment of a genetic ideotype for this region.

References

1. Donald CM (1968) The breeding of crop ideotypes. Euphytica 17:385–403
2. Hodson DP, White JW (2007) Use of spatial analyses for global characterization of wheat-based production systems. J Agric Sci 145: 115–125
3. Eagles HA, Cane K, Kuchel H, Hollamby GJ, Vallance N, Eastwood RF, Gororo NN, Martin PJ (2010) Photoperiod and vernalization gene effects in southern Australian wheat. Crop Pasture Sci 61:721–730
4. Crossa J, Burgueño J, Dreisigacker S, Vargas M, Herrera-Foessel S, Lillemo M, Singh RP, Trethowan R, Warburton M, Franco J, Crouch JH, Ortiz R (2007) Association analysis of historical bread wheat germplasm using additive genetic covariance of relatives and population structure. Genetics 177:1889–1913

5. Rajaram S, van Ginkel M, Fischer RA (1994) CIMMYT's wheat breeding mega-environments (ME). In: Li, Z.S. and Xin, Z.Y. Eds. Proceedings of the 8th International Wheat Genetics Symposium, 19–24 July 1993, Beijing, China. China Agricultural Scientech Press, pp 1101–1106
6. Braun HJ, Rajaram S, Van Ginkel M (1996) CIMMYT's approach to breeding for wide adaptation. Euphytica 92:175–183
7. Hodson DP, Cressman K, Nazari K, Park RF, Yahyaoui A (2009) The global cereal rust monitoring system. In: McIntosh R (ed) Proceedings of the Technical Workshop, Borlaug Global Rust Initiative, Cd. Obregon, Sonora, Mexico, 17–20 March, 2009, pp 35–46
8. Manandhar R, Odeh IOA, Pontius RG (2010) Analysis of twenty years of categorical land transitions in the lower Hunter of New South Wales, Australia. Agric Ecosyst Environ 135:336–346
9. Nelson MA, Odeh IOA, Bishop TFA, Weber N (2010) Quantifying the uncertainty in digital soil class maps developed using model-based approaches. In: Gilkes RJ, Prakonkep N (eds) Proceedings 19th World Congress of Soil Science: Soil solutions for a changing world, Brisbane, Australia. Working Group 1.3 Digital soil assessment pp 42–45
10. Comstock RE (1977) Quantitative genetics and the design of breeding programs. In: Pollack E, Kempthorne O, Bailey B (eds) Proceedings of the international conference on quantitative genetics. Iowa State University Press, Ames, IA, pp 705–718
11. Chenu K, Cooper M, Hammer GL, Mathews KL, Dreccer MF, Chapman SC (2011) Environment characterization as an aid to wheat improvement: interpreting genotype–environment interactions by modelling water-deficit patterns in North-Eastern Australia. J Exp Bot 62:1743–1755
12. Reynolds MP, Trethowan RM (2007) Physiological interventions in breeding for adaptation to abiotic stress, 2007. In: Spiertz JHJ, Struik PC, Van Laar HH (eds) Scale and complexity in plant systems research, gene-plant-crop relations. Springer, The Netherlands, pp 129–146
13. Lopez-Castaneda C, Richards RA, Farquhar GD, Williamson RE (1996) Seed and seedling characteristics contributing to variation in early vigor among temperate cereals. Crop Sci 36: 1257–1266
14. Trethowan RM, Singh RP, Huerta-Espino J, Crossa J, van Ginkel M (2001) Coleoptile length variation of near isogenic Rht lines of modern CIMMYT bread and durum wheat. Field Crop Res 70:167–176
15. Rebetzke GJ, Richards RA, Sirault XRR, Morrison AD (2004) Genetic analysis of coleoptile length and diameter in wheat. Aust J Agric Res 55:733–743
16. Richards RA, Lukacs Z (2002) Seedling vigour in wheat—sources of variation for genetic and agronomic improvement. Aust J Agric Res 53: 41–50
17. GangPing X, McIntyre CL, Jenkins CLD, Glassop D, van Herwaarden AF, Shorter R (2008) Molecular dissection of variation in carbohydrate metabolism related to water-soluble carbohydrate accumulation in stems of wheat. Plant Physiol 146:441–454
18. Nik MM, Babaeian M, Tavassoli A (2011) Effect of seed size and genotype on germination characteristic and seed nutrient content of wheat. Sci Res Essays 6:2019–2025
19. Amani I, Fischer RA, Reynolds MP (1996) Canopy temperature depression associated with yield of irrigated spring wheat cultivars in a hot climate. J Agron Crop Sci 176:119–129
20. Tausz-Posch S, Seneweera S, Norton RM, Fitzgerald GJ, Tausz M (2012) Can a wheat cultivar with high transpiration efficiency maintain its yield advantage over a near-isogenic cultivar under elevated CO_2? Field Crop Res 133: 160–166
21. Damon PM, Ma QF, Rengel Z (2011) Wheat genotypes differ in potassium accumulation and osmotic adjustment under drought stress. Crop Pasture Sci 62:550–555
22. Majdi M, Kamali MRJ, Moghaddam ME, Asli DE, Moradi F, Tahmasbi S (2011) Variation in some agronomic characteristics and soluble stem carbohydrate content at anthesis in spring wheat genotypes under terminal drought stress conditions. Iranian J Crop Sci 13:299–309
23. Izzi G, Farahani HJ, Bruggeman AMD, Oweis TY (2008) In-season wheat root growth and soil water extraction in the Mediterranean environment of northern Syria. Agric Water Manag 95:259–270
24. Dreccer MF, van Herwaarden AF, Chapman SC (2009) Grain number and grain weight in wheat lines contrasting for stem water soluble carbohydrate concentration. Field Crop Res 112:43–54
25. Siahpoosh MR, Dehghanian E (2012) Water use efficiency, transpiration efficiency, and uptake efficiency of wheat during drought. Agron J 104:1238–1243
26. Luo PG, Zhang HY, Shu K, Wu XH, Zhang HQ, Ren ZL (2009) The physiological genetic effects of 1BL/1RS translocated chromosome

in "stay green" wheat cultivar CN17. Can J Plant Sci 89:1–10

27. Duggan BL, Richards RA, van Herwaarden AF, Fettell NA (2005) Agronomic evaluation of a tiller inhibition gene (*tin*) in wheat. I. Effect on yield, yield components, and grain protein. Aust J Agric Res 56:169–178
28. Ibrahim SE, Schubert A, Pillen K, Leon J (2012) QTL analysis of drought tolerance for seedling root morphological traits in an advanced backcross population of spring wheat. Int J Agric Sci 2:619–629
29. Spielmeyer W, Hyles J, Joaquim P, Azanza F, Bonnett D, Ellis ME, Moore C, Richards RA (2007) A QTL on chromosome 6A in bread wheat (Triticum aestivum) is associated with longer coleoptiles, greater seedling vigour and final plant height. Theor Appl Genet 115:59–66
30. Liu X-L, Chang X-P, Li R-Z, Jing R-L (2011) Mapping QTLs for seminal root architecture and coleoptile length in wheat. Acta Agron Sin 37:381–388
31. McIntyre CL, Seung D, Casu RE, Rebetzke GJ, Shorter R, Xue GP (2012) Genotypic variation in the accumulation of water soluble carbohydrates in wheat. Funct Plant Biol 39:560–568
32. Snape JW, Foulkes MJ, Simmonds J, Leverington M, Fish LJ, Wang YK, Ciavarrella M (2007) Dissecting gene * environmental effects on wheat yields via QTL and physiological analysis. Euphytica 154:401–408
33. Rebetzke GJ, Condon AG, Farquhar GD, Appels R, Richards RA (2008) Quantitative trait loci for carbon isotope discrimination are repeatable across environments and wheat mapping populations. Theor Appl Genet 118:123–137
34. Ciuca M, Banica C, David M, Saulescu NN (2010) SSR markers associated with the capacity for osmotic adjustment in wheat (Triticum aestivum L.). Romanian Agric Res 27:1–5
35. Alexander LM, Kirigwi FM, Fritz AK, Fellers JP (2012) Mapping and quantitative trait loci analysis of drought tolerance in a spring wheat population using amplified fragment length polymorphism and diversity array technology markers. Crop Sci 52:253–261
36. Naruoka Y, Sherman JD, Lanning SP, Blake NK, Martin JM, Talbert LE (2012) Genetic analysis of green leaf duration in spring wheat. Crop Sci 52:99–109
37. Paliwal R, Roder MS, Uttam K, Srivastava JP, Joshi AK (2012) QTL mapping of terminal heat tolerance in hexaploid wheat (*T. aestivum* L.). Theor Appl Genet 125:561–575
38. Nezhad KZ, Weber WE, Roder MS, Sharma S, Lohwasser U, Meyer RC, Saal B, Borner A (2012) QTL analysis for thousand-grain weight under terminal drought stress in bread wheat (*Triticum aestivum* L.). Euphytica 186:127–138
39. Kumar S, Sharma V, Chaudhary S, Tyagi A, Mishra P, Priyadarshini A, Singh A (2012) Genetics of flowering time in bread wheat *Triticum aestivum*: complementary interaction between vernalization-insensitive and photoperiod-insensitive mutations imparts very early flowering habit to spring wheat. J Genet 91:33–47
40. Mathews KL, Chapman SC, Trethowan RM, Singh R, Crossa J, Pfeiffer WH, van-Ginkel M, DeLacy I (2006) Global adaptation of spring bread and durum wheat lines near-isogenic for major reduced height genes. Crop Sci 46:603–613
41. Ellis MH, Rebetzke GJ, Azanza F, Richards RA, Spielmeyer W (2005) Molecular mapping of gibberellin-responsive dwarfing genes in bread wheat. Theor Appl Genet 111:423–430
42. Trethowan RM, Crossa J, Pfeiffer WH (2005) Management of genotype × environment interactions and their implications for durum wheat breeding. In: Royo C, Nachit MM, di Fonzo N, Araus JL, Pfeiffer WH, Slafer GA (eds) Durum wheat breeding: current approaches and future strategies. The Harworth Press, Inc., New York, NY, pp 777–802
43. Trethowan RM, Crossa J, van Ginkel M, Rajaram S (2001) Relationships among bread wheat international yield testing locations in dry areas. Crop Sci 41:1461–1469
44. Trethowan RM, van Ginkel M, Ammar K, Crossa J, Payne TS, Cukadar B, Rajaram S, Hernandez E (2003) Associations among twenty years of bread wheat yield evaluation environments. Crop Sci 43:1698–1711
45. Lillemo M, van Ginkel M, Trethowan RM, Hernandez E, Rajaram S (2004) Associations among international CIMMYT bread wheat yield testing locations in high rainfall areas and their implications for wheat breeding. Crop Sci 44:1163–1169
46. Lillemo M, van Ginkel M, Trethowan RM, Hernandez E, Crossa J (2005) Differential adaptation of CIMMYT bread wheat to global high temperature environments. Crop Sci 45:2443–2453
47. Trethowan RM, Crossa J (2007) Lessons learnt from forty years of international bread wheat trials. Euphytica 157:385–390

48. Manès Y, Gomez H, Puhl L, Reynolds M, Trethowan RM (2012) Genetic yield grains of the CIMMYT international semi-arid wheat yield trials from 1994 to 2010. Crop Sci 52:1543–1552
49. Cooper M, Fox PN (1996) Environmental characterization based on probe and reference genotypes. In: Hammer GL, Cooper M (eds) Plant adaptation and crop improvement. CAB International in association with IRRI and ICRISAT, Wallingford, pp 529–547
50. Mathews KL, Trethowan RM, Milgate A, Payne T, van Ginkel M, Crossa J, DeLacy I, Cooper M, Chapman SC (2011) Indirect selection using reference and probe genotype performance in multi-environment trials. Crop Pasture Sci 62:313–327
51. Trethowan RM, Matthews K, Chapman S, Manes Y, Nicol J (2010) An international perspective on breeding for resistance to soil borne pathogens. In: Stirling GR (ed) Proceedings sixth Australian soil-borne diseases symposium. APPS, Toowoomba
52. Vanderplank JE (1963) Plant diseases: epidemics and control. Academic, New York, p 349
53. Sayre KD, Singh RP, Huerta-Espino J, Rajaram S (1998) Genetic progress in reducing losses to leaf rust in CIMMYT-derived Mexican spring wheat cultivars. Crop Sci 38:654–659
54. Eagles HA, Cane K, Vallance N (2009) The flow of alleles of important photoperiod and vernalisation genes through Australian wheat. Crop Pasture Sci 60:646–657
55. Eagles HA, Cane K, Eastwood RF, Hollamby GJ, Kuchel H, Martin PJ, Cornish GB (2006) Contributions of glutenin and puroindoline genes to grain quality traits in southern Australian wheat breeding programs. Aust J Agric Res 57:179–186
56. Gupta PK, Rustgi S, Kulwal PL (2005) Linkage disequilibrium and association studies in higher plants: present status and future prospects. Plant Mol Biol 57:461–485
57. Charmet G, Robert N, Perretant MR, Gay G, Sourdille P, Groos C, Bernard S, Bernard M (1999) Marker-assisted recurrent selection for cumulating additive and interactive QTLs in recombinant inbred lines. Theor Appl Genet 99:1143–1148
58. Cavanagh C, Morell M, Mackay I, Powell W (2008) From mutations to MAGIC: resources for gene discovery, validation and delivery in crop plants. Curr Opin Plant Biol 11:215–221
59. Paux E, Sourdille P, Mackay I, Feuillet C (2012) Sequence-based marker development in wheat: advances and applications to breeding. Biotechnol Adv 30:1071–1088
60. Kuan JZ, Wang SH, Feng YQ, Liu ZX, Wang GX (2006) The 4E- ms system of producing hybrid wheat. Crop Sci 46:250–255
61. DeLacy IH, Fox PN, McLaren G, Trethowan RM, White JW (2009) A conceptual model for describing processes of crop improvement in database structures. Crop Sci 49:2100–2112
62. Moud AAM, Yamagishi T (2006) Differences between water extraction patterns of three wheat (Triticum aestivum L.) cultivars at different soil depths under gradually downward soil drying conditions. J Agric Sci Technol 8:271–279
63. Liu X, Chang X, Li R-Z, Jing R (2011) Mapping QTLs for seminal root architecture and coleoptile length in wheat. Acta Agron Sin 37:381–388
64. Ribaut J-M, de Vicente MC, Delannay X (2010) Molecular breeding in developing countries: challenges and perspectives. Curr Opin plant Biol 13:1–6
65. Delannay X, McLaren G, Ribaut J-M (2012) Fostering molecular breeding in developing countries. Mol Breed 29:857–873

Chapter 2

From Genes to Markers: Exploiting Gene Sequence Information to Develop Tools for Plant Breeding

Melissa Garcia and Diane E. Mather

Abstract

Once the sequence is known for a gene of interest, it is usually possible to design markers to detect polymorphisms within the gene. Such markers can be particularly useful in plant breeding, especially if they detect the causal polymorphism within the gene and are diagnostic of the phenotype. In this chapter, we (1) discuss how gene sequences are obtained and aligned and how polymorphic sites can be identified or predicted; (2) explain the principles of PCR primer design and PCR amplification and provide guidelines for their application in the design and testing of markers; (3) discuss detection methods for presence/absence (dominant) polymorphisms, length polymorphisms and single nucleotide polymorphisms (SNPs); and (4) outline some of the factors that affect the utility of markers in plant breeding and explain how markers can be evaluated (validated) for use in plant breeding.

Key words Molecular markers, Plant breeding, Marker-assisted selection, Functional markers

1 Introduction

In plant breeding, gene-based markers can be used to directly track and select for (or against) particular alleles. Ideally these markers assay the functional DNA polymorphisms within genes, providing "perfect" markers that are diagnostic across all germplasm. Markers can also be designed to detect polymorphisms that have no known function; such markers may be used in cases where the functional polymorphism is difficult to assay or has not been identified. This chapter discusses the development of marker assays for individual genes for which sequence information is available from at least one source. The workflow is described in Fig. 1.

Delphine Fleury and Ryan Whitford (eds.), *Crop Breeding: Methods and Protocols*, Methods in Molecular Biology, vol. 1145, DOI 10.1007/978-1-4939-0446-4_2, © Springer Science+Business Media New York 2014

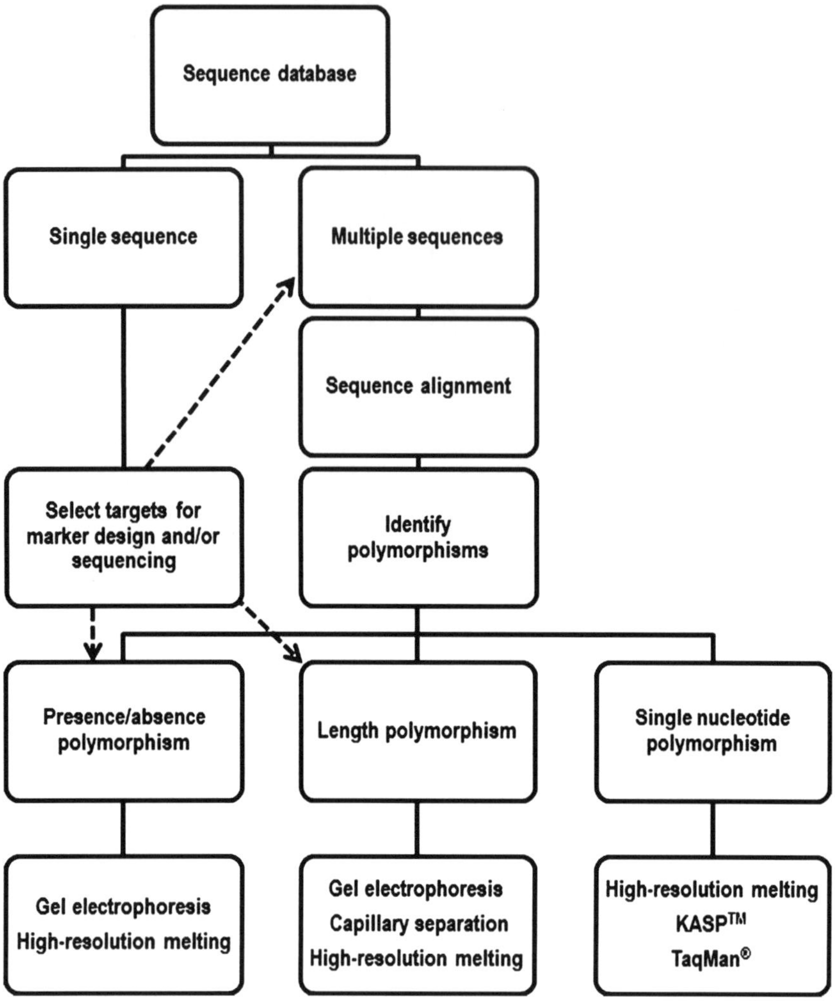

Fig. 1 Schematic representation pathways from a database search to detection of the polymorphism

2 Obtaining Sequences from Databases

The first step in designing markers for a particular gene is to assemble and understand the available sequence information for that gene. This should involve (1) a database search to obtain all available sequence information for the gene and (2) a literature search aimed at finding out whether any specific polymorphisms have already been demonstrated to have functional effects.

The three most important sequence databases are GenBank (maintained by the National Center for Biotechnology Information or NCBI, http://www.ncbi.nlm.nih.gov/genbank/), the European Molecular Biology Laboratory Nucleotide Sequence Database (EMBL-Bank, http://www.ebi.ac.uk/embl/), and the

DNA Database of Japan (DDBJ, http://www.ddbj.nig.ac.jp/). These databases collaborate and exchange updated information on sequences collected worldwide. Each sequence deposited in one of these databases is assigned a unique accession number that is common to all databases. That accession number should be given in any published paper describing the sequence. The sequence of a published gene can be obtained from these databases by performing a search using the accession number, the name of the gene or protein or related key words. It is always preferable to perform the search with the accession number because of its uniqueness.

When sequence is available from just one source (e.g., one cultivar), that sequence can be used to identify regions that are likely to be polymorphic and (or) in which polymorphisms could be readily detected. Primers can be designed to amplify these regions from other sources (plants, lines, cultivars or accessions). The resulting amplicons can be directly assayed for polymorphisms using various marker technologies, or can be sequenced to determine whether there is polymorphism among sources of interest [1, 2].

If the gene of interest has been more extensively studied and a specific polymorphism has been demonstrated to have a functional effect, that polymorphism will likely be chosen as the priority target for marker design. If sequences are already available from more than one source, polymorphisms can be detected in silico by aligning those sequences. Alignments do not need to be restricted to complete gene sequences that have already been annotated as belonging to the gene of interest; a known gene sequence can be used as the query in BLAST [3] searches to retrieve highly similar sequences from databases.

3 Identifying Target Sites Within Individual Sequences

Sequence information from a single source can be used to find potential target sites for marker design. These may include restriction sites and the positions of features that are likely to be polymorphic across sources, such as introns, simple sequence repeats (SSR), and insertion site-based polymorphisms (ISBP).

3.1 Restriction Sites

Restriction sites are specific nucleotide sequences that are recognized and cleaved by restriction enzymes. Since even a single nucleotide change can create or destroy such a site, DNA polymorphisms can affect whether or how many times the restriction enzyme will cleave a DNA fragment or amplicon. This can lead to length polymorphisms after DNA is incubated with the restriction enzyme. Historically, this was widely exploited for marker design using restriction fragment length polymorphism [4–6]. It has also been used in combination with the polymerase chain reaction (PCR). For example, cleaved amplified polymorphic sequence

(CAPS) markers [7] can reveal polymorphism at restriction sites using PCR amplification with primers that flank one or more restriction sites followed by digestion of the PCR product with the appropriate restriction enzyme [8, 9]. Webcutter (http://rna.lundberg.gu.se/cutter2/) and RestrictionMapper (http://www.restrictionmapper.org/) are web-based programs designed to identify restriction sites.

3.2 Introns

Introns are generally expected to harbor more sequence variation than exons. This is because sequence changes within introns are less likely to affect the structure and function of gene products than sequence changes within exons [10]. Intron-related polymorphisms that can be targeted for marker design include presence/absence of a particular intron, differences in intron length and sequence polymorphism within the intron. If the gene sequence has been annotated and information is available on the positions of introns, primers can be designed to amplify introns and their flanking sequences. These primers are usually anchored in the conserved sequence of the exons that flank the target intron. If the positions of the introns are not yet known, they can be predicted by aligning the genomic sequence with cDNA sequences or by using programs that predict exon-intron boundaries (e.g., FGENESH [11], http://linux1.softberry.com).

3.3 Simple Sequence Repeats

Simple sequence repeats are sequence motifs of one or a few nucleotides that are repeated in tandem [12]. The number of times the sequence motif is repeated can vary among individuals, providing a length polymorphism when a product is amplified using primers that flank the repeats [13, 14]. There are many available programs for identifying SSRs, including SciRoKo [15] (http://kofler.or.at/bioinformatics/SciRoKo/) and SSRIT [16] (http://www.gramene.org/db/markers/ssrtool). If only one gene sequence is being analyzed, SSRs can be identified manually.

3.4 Insertion Site-Based Polymorphisms

Insertions of transposable elements (TE) are frequent in plant genomes [17], providing a rich source of DNA polymorphism. TE-related polymorphisms include presence/absence of the insertion, difference in the length of the insertion and sequence polymorphism within the insertion. Insertion of a TE can create a unique junction between the TE and the TE-harboring DNA sequence. With the software IsbpFinder.pl, it is possible to automatically detect these junctions and to design primer pairs for the development of ISBP markers [18]. Within each pair, one primer is designed in the TE and the other in one of the flanking DNA regions. Primers can be designed for any DNA sequence and then evaluated in potentially polymorphic materials.

4 Multiple Sequence Alignment and Identification of Polymorphisms

Where sequence is available from more than one source, multiple sequences can be aligned to identify the polymorphisms. Both commercial and noncommercial bioinformatic tools are available for sequence alignment. These include Clustal [19] and Mafft [20], both of which can be downloaded or used online at the European Bioinformatics Institute (EBI) Web site (http://www.ebi.ac.uk/Tools/msa/). Software is also available for the identification of polymorphisms [21, 22], but this process is usually done manually if only one or a few genes are to be analyzed.

Multiple sequence alignment can reveal different types of DNA polymorphism, including simple sequence repeats (SSR), single nucleotide polymorphisms (SNP), and small insertions and deletions (indels) (Fig. 2). Single nucleotide polymorphisms, which involve differences at individual nucleotide sites, are the most common type of sequence polymorphism. At a particular SNP site, there are usually only two alleles and there are only six possible combinations: the pyrimidine–pyrimidine combination cytosine–thymine (C/T), the purine–purine combination adenine–guanine (A/G) and four pyrimidine/purine combinations: C/A, T/A, C/G and T/G.

Figure 2 shows the alignment of partial DNA sequence from the same hypothetical gene from three cultivars (1 to 3). The polymorphisms in this sequence alignment include a polymorphic SSR consisting of GA repeat (A), an indel (B) where cultivars 1 and 2 have an extra AC relative to cultivar 3 and a T/A SNP (C) that distinguishes cultivar 2 from cultivars 1 and 3. The choice of which polymorphism to target for marker design would depend on previous knowledge about the sequence and on how easy it is to assay the polymorphism. For the design of markers that will be broadly

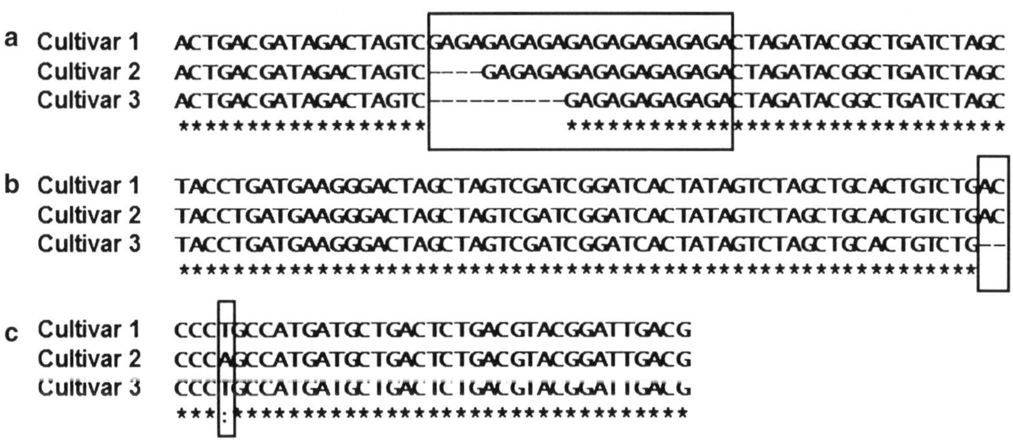

Fig. 2 Alignment of three DNA sequences showing three types of polymorphism: (**a**) simple sequence repeat (SSR); (**b**) insertion deletion (indel); and (**c**) single nucleotide polymorphism (SNP) Melissa Garcia and Diane E. Mather

applicable across germplasm, it is best to use polymorphisms near (or preferably at) the functional site, as they can be expected to remain associated with trait differences across germplasm. In some cases, however, the functional polymorphism may not be known or cannot be assayed because its harboring sequence is not suitable for primer design (*see* Subheading 5.1 of this chapter). In such cases, any other polymorphisms within the same gene (or adjacent sequence) can be considered as possible targets, as they all would be expected to co-segregate with the functional polymorphism in breeding populations.

5 DNA Amplification by the Polymerase Chain Reaction

Most marker technologies that are used in plant breeding employ the polymerase chain reaction (PCR) to amplify polymorphic segments of DNA. This increases the quantity of the target DNA to facilitate detection of the polymorphism. Understanding PCR and the principles of PCR primer design is important for designing and testing gene-based markers.

5.1 PCR Primer Design

Methods for primer design are well established [23, 24] and many computer programs are available that implement these. Among these, Primer 3 [25–27] is one of the most popular because of its capabilities and online accessibility. Regardless of the program used, the key factors to be considered are primer length, melting temperature (T_m), secondary structures, and GC content.

- *Primer length*: The length of a primer influences its specificity, T_m and time required for annealing. In general, the longer the primer, the higher the primer specificity, but primers that are too long have higher probability of forming secondary structures such as hairpins and dimers. Primer lengths ranging from 18 to 28 bp are considered optimal for most PCR applications.
- *Primer melting temperature*: The T_m of a primer depends on its length and its base composition. For most PCR applications, the primer T_m should be between 55 and 65 °C and should not differ by more than 3 °C between the two members of a primer pair.
- *Secondary structures*: Primers whose sequences have regions of self-homology may form self-dimers or fold to form hairpin structures. Similarly, cross-dimers can be formed by annealing between primers. These secondary structures compromise the PCR by reducing the availability of primers. The likelihood of secondary structures forming spontaneously can be assessed as ΔG; as a rule of thumb, the ΔG of hairpins should not be lower than −1 kcal/mol and the ΔG of the dimers should not be lower than −4 kcal/mol.

- *GC content*: Because G–C bonds are stronger than A–T bonds, the stability of annealing between a primer and its target depends largely on the GC content (number of Gs and Cs in the primer as a percentage of the total number of base pairs) of the primer. The GC content also increases the primer T_m. A primer should normally have a GC content between 40 and 60 %, and should preferably have Gs or Cs within a few bases of its 3′ end to increase its specificity.
- *Locus specificity*: Primers should be designed to amplify a single region of the genome. Primer sequences can be used to perform a BLAST search. If any hit other than the target sequence is detected, new primers should be designed to avoid non-target amplification.

Primer design for polymorphism detection usually involves either the design of a pair of primers that flank the polymorphic region or a pair of primers in which one primer is at the polymorphic site. The choice between these depends on the type of polymorphism to be assayed and the method that will be used for polymorphism detection.

Designing primers to flank a polymorphic site: each of the primers designed to flank a polymorphic sites should anneal to a region of sequence conservation, as this will allow the use of the same primers across diverse genotypes. When multiple primer sites are available, pairs of sites are chosen at positions that will provide amplicons with lengths suitable for the detection method to be used (*see* Subheading 6 of this chapter).

Designing primers at a polymorphic site: Some SNP marker assays make use of an allele-specific primer designed to have the base at its 3′ end complementary to one of the SNP alleles. The specificity of amplification using such a primer relies upon stringent conditions under which a mismatch at the 3′ end prevents annealing of the primer and amplification of the alternative allele. For pyrimidine–pyrimidine (C/T) SNPs and purine–purine (A/G), a single mismatch is likely to be sufficient to destabilize primer annealing and avoid amplification of the alternative allele. For pyrimidine–purine (C/A, T/A, C/G, or T/G) SNPs, it may be necessary to include an additional artificial mismatch near the 3′ end of the primer to improve specificity of amplification.

5.2 Synthesis of Primers for Use in PCR

After primer design, the sequences of the primers are sent for primer synthesis. Many companies provide this service, with the price usually depending on the length of the primer, any special features of the primer, the quantity of primer ordered, the number of primers ordered, and the container(s) in which the primers are to be delivered. When large numbers of primers are ordered, they can be delivered in microwell plates; this is usually less expensive than delivery in individual tubes.

For some genotyping technologies, primers need to be labelled with fluorescent tags. This adds considerable cost to the primer synthesis, but it enables the fluorescent tag to be incorporated into amplicons during PCR, allowing their detection.

5.3 PCR Amplification

A typical PCR reaction includes PCR buffer, deoxynucleotide triphosphates (dNTP), magnesium, thermostable DNA polymerase, forward and reverse primers, and target DNA.

- *PCR buffer*: The role of the PCR buffer is to maintain the pH in an optimum range for PCR. The optimum pH varies among DNA polymerases. The buffer, which usually contains KCl and Tris–HCl, is generally provided as part of a DNA polymerase kit.
- *Magnesium*: Magnesium is an essential cofactor for the DNA polymerase, but at high concentrations it can reduce the fidelity of the polymerase. Magnesium may be included in the PCR buffer or supplied separately in the form of magnesium chloride, so that the magnesium concentration can be adjusted to optimize target amplification. Concentrations between 1.5 mM and 2.5 mM are optimum for most amplification reactions.
- *dNTPs*: The four DNA deoxynucleotides (A, T, C, and G) required by the DNA polymerase to synthesize new DNA strand are provided as dNTP. Each dNTP should be included in equimolar concentration in the PCR. This concentration is usually between 50 and 200 μM. Excessive dNTP concentration can inhibit the synthesis of DNA and/or cause misincorporation of nucleotides.
- *Thermostable DNA polymerase*: Thermostable DNA polymerases are able to catalyze the synthesis of DNA even at the high temperatures used in the PCR. Initial tests are usually necessary to determine the concentration of DNA polymerase to be used. If too little polymerase is used, primer extension can be incomplete or fail to occur. If too much polymerase is used, there can be amplification of nonspecific PCR products. The standard DNA polymerase is Taq polymerase [28, 29], but many other DNA polymerases have been developed [30] to overcome some of the limitations of Taq such as the lack of a 3′→5′ exonuclease proofreading activity [31, 32].
- *Primers*: DNA polymerases add nucleotides to the 3′ ends of DNA strands, starting with the 3′ ends of the PCR primers. Each PCR primer is usually included at a concentration between 0.1 and 1.0 μM. Higher concentrations can cause nonspecific primer binding and increase the likelihood of primer-dimer formation.
- *Template DNA*: For use in PCR, the genomic DNA from the plants to be assayed should be intact, sufficiently pure and present at an appropriate concentration. Prior to PCR, the integrity

of the DNA can be tested by electrophoresis of an aliquot of extracted DNA on agarose gel. Intact DNA migrates as a single band of high molecular weight, while degraded DNA results in a smear of fragments. Depending on the DNA isolation method used, DNA samples may contain residues of chemicals used during the extraction process, such as ionic detergents, phenol, isopropanol, ethylenediaminetetraacetic acid (EDTA), or ethanol. Although such residues can inhibit PCR, some degree of contamination may be tolerated in order to reduce the cost and increase the throughput of DNA extraction and purification. Usually, initial tests are conducted to determine the optimum amount to DNA to include. Generally, this is between 5 and 100 ng in a 25 μl reaction. High DNA concentration can cause PCR failure by interfering with the access of the polymerase to the DNA molecule. If the DNA used in the PCR is not purified, an increase of the DNA concentration in the PCR will also increase the concentrations of residues that can inhibit the PCR.

After preparing the PCR mix, cycling conditions need to be optimized. A typical PCR program involves an initial denaturation of the DNA at 95 °C in order to separate the DNA strands, followed by between 25 and 40 cycles involving denaturation at 95 °C, primer annealing at between 55 and 60 °C and primer extension at 72 °C.

Most of the optimization in the PCR cycling conditions is related to establishing the best annealing temperature (T_a). This usually starts with a T_a that is 3–5 °C below the T_m calculated during primer design. In practice, primers with T_m between 55 and 65 °C usually work at a T_a of 60 °C. The T_a determines whether the primer will anneal only to a perfectly matched sequence or also to sequences with a few mismatches. There are two situations in which adjustments of the T_a might improve the PCR results:

- *PCR failure or weak amplification*: If PCR fails or the PCR product is very weak, a lower T_a should be tested. It is important to keep in mind that lowering the T_a can cause the primers to anneal to nonspecific regions. Usually, annealing temperatures below 55 °C should be avoided.
- *Presence of nonspecific amplification*: Where products other than the target sequence are amplified, an increase in the T_a might increase primer specificity and avoid nonspecific amplification.

Annealing temperatures should be adjusted gradually; increases or decreases of 2 °C are usually sufficient to affect the results.

The number of cycles and the duration of each step depend on the type of PCR being performed. Increasing the number of cycles can improve the yield of the target, but can also increase the yield of nonspecific products. For most PCR applications, 35 cycles are sufficient and the use of more than 40 cycles should be avoided. The initial denaturation of the DNA should be done for between

1 and 3 min and during the cycling this step should be reduced to around 30 s to avoid decreasing the polymerase activity. The duration of the primer annealing step depends on the length of the primer but usually 15–30 s are sufficient for the primer to anneal. The duration of the extension step depends on the DNA polymerase used and the size of the product being amplified. Although Taq polymerase can extend up to 100 bp per second, it is usually accepted that 1 min is needed to extend 1,000 bp.

Touchdown PCR [33] is a method that can improve both specificity and amplification. In touchdown PCR, the T_a starts at 65 °C, which is above the primer T_m and is decreased by 1 °C every two cycles until it "touches down" at 55 °C, with 10 further cycles carried out at this T_a. The initial high T_a favors specific amplification of the target sequence and avoids amplification of other sequences. Starting at a high T_a is intended to ensure that the target sequence itself is the first one to be amplified, giving it a "head start" over nonspecific sequences. During later cycles, specific amplicons from the initial cycles will serve as template for further amplification, outcompeting any nonspecific sequences even at low temperatures at which nonspecific annealing might otherwise occur. Touchdown PCR has shown to be very flexible and it is often used as a standard PCR program.

Usually, each PCR is prepared to amplify a single target sequence, but when a large number of regions are to be assayed, it can be useful to design all primer pairs to have similar T_m, allowing the use of the same cycling conditions to amplify different regions. It can also be useful to design assays that amplify multiple target sequences in a single PCR reaction (multiplex PCR [34]). Multiplex PCR is more difficult than uniplex PCR. It requires primers with similar T_a and that will not interact significantly with each other. Optimization of a multiplex reaction usually involves adjustment of both primer concentrations and PCR conditions and comparison of results to those obtained with uniplex PCR. If multiple PCR products are to be distinguished using gel electrophoresis, the sizes of the products generated by different primer pairs need to be different enough to allow their separation. In methods that use detection of fluorescence, different primers may need to be labelled with fluorescent tags that have different emission wavelengths [35].

After PCR is performed, amplicons should be separated by agarose gel electrophoresis to confirm that the sizes of products are as expected. If product sizes differ from expectations, further optimization of PCR conditions may be required.

6 Marker Types and Detection Methods

Markers can be classified based on their ability to distinguish between heterozygotes and homozygotes. A codominant marker is able to distinguish alternative homozygous genotypes from each other and distinguish the heterozygotes from each homozygote. A dominant marker detects just the presence or absence of one particular allele; it cannot distinguish the heterozygotes from the homozygote that carries that allele.

6.1 Detection of Presence/Absence Polymorphisms (Dominant Markers)

The detection of presence/absence polymorphisms is usually done by agarose gel electrophoresis, with the gel concentration adjusted according to the size of the amplicons. Fragments shorter than 50 bp are usually avoided. Fragments ranging from about 80 bp to several Kb may be separated using agarose concentrations between 0.8 % and 3 %. Fluorescent DNA dyes such as ethidium bromide, SYBR® Green (Invitrogen) and GelRed™ (Biotium) are added to the gel and the PCR products can be visualized under UV light. Where dominant markers are used, PCR failure can lead to genotyping errors. If no product is observed, one cannot be certain whether the allele is absent or the PCR has failed. In some cases, a complementary assay can be designed to amplify the alternative allele and it may be possible to optimize two complementary assays to work together in a single PCR reaction. A pair of complementary dominant assays is as informative as a codominant assay because it permits the identification of heterozygotes.

6.2 Detection of Length Polymorphisms

Size differences, such as the ones obtained with SSR and CAPS markers, may be detectable using agarose electrophoresis polyacrylamide gel electrophoresis (PAGE), capillary separation, or high-resolution melting (HRM). The choice among these detection methods depends partly on the size of the difference that needs to be detected.

Agarose gels can be used to resolve fragments that differ by at least 5 bp, while PAGE, capillary electrophoresis and HRM may allow the separation of fragments that differ in length by as little as 1 bp. PAGE can be used to separate PCR products that are as short as about 10 bp up to those that are about 1,000 bp in length. For DNA separation, the polyacrylamide concentration is usually between 3 % and 8 %. The detection of the PCR products can be done by different methods, including the staining with ethidium bromide, SYBR® Green, or silver nitrate. Some systems used for PAGE can detect fluorescently labelled amplicons, overcoming the need for gel staining.

Capillary separation allows automated detection of fluorescently labelled amplicons as they migrate through a capillary. The fluorescence signals are converted into digital data that is compatible with the software used for analysis. Capillary electrophoresis is generally used for products ranging from 120 to 1,200 bp, making it particularly suitable for applications in which markers with different product sizes can be multiplexed together.

High-resolution melting technology [36, 37] is based on detecting differences in the melting behavior of PCR products.

Target DNA sequences are amplified in the presence of a dye that fluoresces when intercalated with double-stranded DNA. The temperature is gradually raised and the PCR products melt (separate into single-stranded DNA), releasing the fluorophore. During the melting of the PCR products, fluorescence levels are recorded to obtain a melting curve for each sample. Amplicons of different sizes generally have different melting curves, allowing the genotyping of length polymorphisms. Amplicons between 80 bp and 250 bp long are preferable for HRM. With larger amplicons, small differences in length may not detectably alter the shapes of melting curves.

6.3 Detection of SNPs

As described above, digestion with restriction enzymes can make it possible to generate length polymorphisms based on SNPs at restriction sites (e.g., in CAPS markers). Although some CAPS markers have been used in plant breeding, they are not very convenient because they require both PCR and enzymatic digestion. Further, this approach is applicable only for SNPs that destroy or create a restriction site. With appropriate primer design and PCR protocols, it is sometimes possible to generate amplicons of different lengths based on differences at just one nucleotide (e.g., temperature-switch PCR [38]). Various other methods are available to detect SNPs without relying on length polymorphisms. These include HRM, TaqMan®, and KASP™.

When HRM is used to distinguish between alleles that differ only by a SNP, primers are usually designed to amplify quite small products (less than about 150 bp), with the polymorphic site near the middle of the product [39, 40]. A single SNP within such a product may or may not modify the melting curve sufficiently for alternative alleles to be differentiated. If it does, the resulting marker may be codominant (with three distinct curves for the heterozygote and the two homozygotes) or dominant (with the heterozygote distinguishable from one homozygote but not the other) or it may distinguish the heterozygote from homozygotes but not the homozygotes from each other.

TaqMan® genotyping assays [41, 42] use PCR primers in combination with a dual-labelled allele-specific probe. The probe contains a fluorophore at its 5′ end and a quencher at its 3′ end. When the probe is intact, the quencher is close enough to the fluorophore to reduce the emission of fluorescence by fluorescence resonance energy transfer (FRET). During PCR, the probe binds specifically to the target site between the primers. During primer extension, the 5′ exonuclease activity of the polymerase degrades the probe, releasing the fluorophore from the quencher. Fluorescence is emitted and can be detected with real-time PCR. Probes and primers for Taqman® assays can be designed using software provided by commercial suppliers of the probes. The probes are usually between 20 and 30 bp long, providing specificity, with the target polymorphism as close as possible to the center of the probe. The primers are designed to closely flank the probe target sequence, to amplify a product no longer than 300 bp.

KASP SNP genotyping (LGC Genomics, UK) is based on allele-specific amplification and FRET. Each assay involves three unlabelled primers: two allele-specific primers and a common primer. The two allele-specific primers (ASP) are designed with their 3′ ends complementary to each of the SNP alleles and with a noncomplementary tail in the 5′ end. Their tail sequences differ from each other and are complementary to two FRET quenching reporter oligonucleotides (each labelled with a different fluorophore) that are present in the KASP Mastermix. During PCR, the fluorescent oligonucleotides are incorporated to the PCR products. The resulting fluorescent products can be detected using a plate reader or real-time PCR instrument. LGC Genomics (http://www.lgcgenomics.com/) provides services for design, validation, and genotyping of KASP assays and reagents for in-house genotyping of KASP markers.

7 Evaluation of the Utility of a Marker for Selection in Plant Breeding

For a marker to be useful for selection in plant breeding, it needs to be easy to assay on a routine basis, robust enough to produce reliable results and diagnostic across breeding germplasm. For application of a marker in multiple breeding programs and facilities, it helps if the marker is versatile enough to allow the use of different genotyping technologies.

A marker that assays the functional polymorphism within a gene can usually be expected to be diagnostic across any germplasm. In contrast, a marker that assays some other polymorphism within the gene may or may not be associated with the trait across all material. Often, the first step in validating a marker is to assay it on the parents of a population that segregates for the trait and for which phenotypic data are available. If the marker is polymorphic between the parents, it is then assayed on the members of the population and the strength of the association of the marker genotype with the phenotype is evaluated. Significant association indicates the potential utility of the marker for breeding. Lack of association may mean that the marker detects a polymorphism in a region of the genome other than the one that directly affects the trait. This can occur even in cases where the marker has been designed based on sequence information for the correct gene, especially in polyploid species, in which there may be several homeologous copies of genes.

Even if a new marker has been confirmed to be effective in one or more mapping populations, it still needs to be validated in a wider range of germplasm. This normally involves assaying the marker on a panel of cultivars and/or breeding material for which reliable phenotypic data are available. The members of the panel should be chosen to represent the diversity of material used in breeding programs. In cases where a marker does not validate perfectly across such a panel, pedigree information can be useful for interpretation of the results.

8 Conclusion

The abundance of plant genomic sequence that can now be generated at low cost, in combination with advances in knowledge about the functions of specific genes, creates opportunities for the design of useful markers based on DNA polymorphisms within gene sequences. Numerous methods are available to exploit these sequences for marker development. Markers used in plant breeding are assayed across large numbers of individual plants and must be easy to assay at low cost. Accuracy and reliability are important, but as molecular marker information is only one of the many factors supporting breeding decisions, occasional errors and assay failures can usually be tolerated. Gel electrophoresis methods are easy and do not require sophisticated equipment, but they can be impractical for the analysis of thousands of samples. Markers that require additional steps, such as restriction digestion, are usually avoided in plant breeding because of the time required for sample preparation before the polymorphism is detected [43]. Simple closed-tube assays are generally preferred, because they are easy to prepare and quickly assayed with little risk of cross-contamination. When only one or a few markers are implemented in a breeding program, or when the combination of markers used changes frequently (among breeding populations and/or as new markers become available), individual-gene (uniplex) marker assays provide the most flexibility. As increasing numbers of markers become available for routine use across breeding germplasm, efficiencies can be gained by combining those markers into multiplexed assays or arrays, allowing for simultaneous assays of multiple genes on large numbers of selection units.

References

1. Ganal MW, Altmann T, Röder MS (2009) SNP identification in crop plants. Curr Opin Plant Biol 12:211–217
2. Rafalski A (2002) Applications of single nucleotide polymorphisms in crop genetics. Curr Opin Plant Biol 5:94–100
3. Altschul SF, Gish W, Miller W et al (1990) Basic local alignment search tool. J Mol Biol 215:403–410
4. Botstein D, White RL, Skolnick M et al (1980) Construction of a genetic linkage map in man using restriction fragment length polymorphisms. Am J Hum Genet 32:314–331
5. Tanksley SD, Young ND, Paterson AH et al (1989) RFLP mapping in plant breeding: new tools for an old science. Nature 7:257–264
6. Powell W, Morgante M, Andre C et al (1996) The comparison of RFLP, RAPD, AFLP and SSR (microsatellite) markers for germplasm analysis. Mol Breed 2:225–238
7. Konieczny F, Ausubel A (1993) A procedure for mapping *Arabidopsis* mutations using

co-dominant ecotype-specific PCR-based markers. Plant J 4:403–410

8. Baumbusch LO, Sundal INAK, Hughes DW et al (2001) Efficient protocols for CAPS-based mapping in *Arabidopsis*. Plant Mol Biol Rep 19:137–149
9. Agarwal M, Shrivastava N, Padh H (2008) Advances in molecular marker techniques and their applications in plant sciences. Plant Cell Rep 27:617–631
10. Holland JB, Helland SJ, Sharopova N et al (2001) Polymorphism of PCR-based markers targeting exons, introns, promoter regions, and SSRs in maize and introns and repeat sequences in oat. Genome 44:1065–1076
11. Salamov AA, Solovyev VV (2000) Ab initio gene finding in *Drosophila* genomic DNA. Genome Res 10:516–522
12. Tautz D, Renz M (1984) Simple sequences are ubiquitous repetitive components of eukaryotic genomes. Nucleic Acids Res 12:4127–4138
13. Litt M, Luty JA (1989) A hypervariable microsatellite revealed by *in vitro* amplification of dinucleotide repeat within the cardiac muscle actin gene. Am J Hum Genet 44:398–401
14. Li Y-C, Korol AB, Fahima T et al (2002) Microsatellites: genomic distribution, putative functions and mutational mechanisms: a review. Mol Ecol 11:2453–2465
15. Kofler R, Schlötterer C, Lelley T (2007) SciRoKo: a new tool for whole genome microsatellite search and investigation. Bioinformatics 23:1683–1685
16. Temnykh S, DeClerck G, Lukashova A et al (2001) Computational and experimental analysis of microsatellites in rice (*Oryza sativa* L.): frequency, length variation, transposon associations, and genetic marker potential. Genome Res 11:1441–1452
17. Feschotte C, Jiang N, Wessler SR (2002) Plant transposable elements: where genetics meets genomics. Nat Rev Genet 3:329–341
18. Paux E, Faure S, Choulet F et al (2010) Insertion site-based polymorphism markers open new perspectives for genome saturation and marker-assisted selection in wheat. Plant Biotechnol J 8:196–210
19. Larkin HD, Blackshields MA, Brown G et al (2007) Clustal W and Clustal X version 2.0. Bioinformatics 23:2947–2948
20. Katoh M, Kuma M (2002) MAFFT: a novel method for rapid multiple sequence alignment based on fast Fourier transform. Nucleic Acids Res 30:3059–3066
21. Marth GT, Korf I, Yandell MD et al (1999) A general approach to single-nucleotide polymorphism discovery. Nat Genet 23:452–456
22. Barker G, Batley J, O' Sullivan H et al (2003) Redundancy based detection of sequence polymorphisms in expressed sequence tag data using autoSNP. Bioinformatics 19:421–422
23. Apte A, Daniel S (2009) PCR primer design. Cold Spring Harb Protoc. doi:10.1101/pdb.ip65
24. Dieffenbach GS, Dveksler CW (1995) PCR primer: a laboratory manual. Cold Spring Harbor Laboratory Press, Cold Spring Harbor
25. Koressaar T, Remm M (2007) Enhancements and modifications of primer design program Primer3. Bioinformatics 23:1289–1291
26. Untergasser LJ, Nijveen A, Rao H et al (2007) Primer3Plus, an enhanced web interface to Primer3. Nucleic Acids Res 35:W71–W74
27. Rozen S, Skaletsky H (2000) Primer3 on the WWW for general users and for biologist programmers. In: Krawetz S, Misener S (eds) Bioinformatics methods and protocols: methods in molecular biology. Humana Press, Totowa, NJ, pp 365–386
28. Saiki RK, Gelfand DH, Stoffel S et al (1988) Primer-directed enzymatic amplification of DNA with a thermostable DNA polymerase. Science 239:487–491
29. Chien A, Edgar DB, Trela JM (1976) Deoxyribonucleic acid polymerase from the extreme thermophile *Thermus aquaticus*. J Bacteriol 127:1550–1557
30. Kolmodin LA, Birch DE (2002) Polymerase chain reaction: basic principles and routine practice. In: Chen BY, Janes HW (eds) PCR cloning protocols. Humana Press, Totowa, NJ, pp 3–18
31. Eckert KA, Kunkel TA (1991) DNA polymerase fidelity and the polymerase chain reaction. PCR Methods Appl 1:17–24
32. Tindall KR, Kunkel TA (1988) Fidelity of DNA synthesis by the *Thermus aquaticus* DNA polymerase. Biochemistry 27:6008–6013
33. Don RH, Cox PT, Wainwright BJ et al (1991) "Touchdown" PCR to circumvent spurious priming during gene amplification. Nucleic Acids Res 19:4008
34. Henegariu O, Heerema NA, Dlouhy SR et al (1997) Multiplex PCR : critical parameters and step-by-step protocol. Biotechniques 23: 504–511
35. Hayden MJ, Nguyen TM, Waterman A et al (2007) Application of multiplex-ready PCR for fluorescence-based SSR genotyping in barley and wheat. Mol Breed 21:271–281
36. Wittwer CT, Reed GH, Gundry CN et al (2003) High-resolution genotyping by amplicon melting analysis using LCGreen. Clin Chem 49:853–860
37. Herrmann H, Durtschi J, Wittwer C et al (2007) Expanded instrument comparison of

amplicon DNA melting analysis for mutation scanning and genotyping. Clin Chem 53: 1544–1548

38. Tabone T, Mather DE, Hayden MJ (2009) Temperature switch PCR (TSP): Robust assay design for reliable amplification and genotyping of SNPs. BMC Genomics 10:580
39. Studer B, Jensen LB, Fiil A et al (2009) "Blind" mapping of genic DNA sequence polymorphisms in *Lolium perenne* L. by high resolution melting curve analysis. Mol Breed 24: 191–199
40. Dong C, Vincent K, Sharp P (2009) Simultaneous mutation detection of three homoeologous genes in wheat by High Resolution Melting analysis and Mutation Surveyor®. BMC Plant Biol 9:143
41. Lee LG, Connell CR, Bloch W et al (1993) Allelic discrimination by nick-translation PCR with fluorogenic probes. Nucleic Acids Res 21: 3761–3766
42. Holland PM, Abramson RD, Watson R et al (1991) Detection of specific polymerase chain reaction product by utilizing the 5′→3′ exonuclease activity of *Thermus aquaticus* DNA polymerase. Proc Natl Acad Sci U S A 88: 7276–7280
43. Bagge M, Lübberstedt T (2008) Functional markers in wheat: technical and economic aspects. Mol Breed 22:319–328

Chapter 3

Temperature Switch PCR (TSP): A Gel-Based Molecular Marker Technique for Investigating Single Nucleotide Polymorphisms

Le Phuoc Thanh and Kelvin Khoo

Abstract

Temperature Switch PCR (TSP) is a robust single-marker single nucleotide polymorphism (SNP) genotyping technique with broad applications in genetic studies of various organisms. The technique consists of a biphasic PCR with two sets of primers, a locus-specific set and a nested locus-specific set. The PCR products can be easily assessed for polymorphism based on different band sizes using agarose gel electrophoresis.

Key words TSP, SNP, Locus-specific primers, Gel electrophoresis, Molecular marker

1 Introduction

Over the past two decades, molecular markers have been utilized with great success in plant breeding [1–8]. More recently, new techniques have significantly increased the efficiency of DNA polymorphism screening compared to conventional methods [4, 9, 10]. Depending on the DNA sequence and the nature of the target polymorphism of interest as well as the structure of the genome surrounding the region, a variety of molecular markers such as Simple Sequence Repeats (SSR), Cleaved Amplified Polymorphic Sequences (CAPS), Inter Simple Sequence Repeats (ISSR), Random Amplified Polymorphic DNAs (RAPD), Amplified Fragment Length Polymorphisms (AFLP), Restriction Fragment Length Polymorphisms (RFLP), Sequenced Characterized Amplified Region Markers (SCAR), Sequence-Tagged Sites (STS), Insertion Site-Based Polymorphisms (ISBP) could be used to detect polymorphisms between different genotypes [2, 3, 11]. PCR products generated from the above mentioned assays are detected by either

Delphine Fleury and Ryan Whitford (eds.), *Crop Breeding: Methods and Protocols*, Methods in Molecular Biology, vol. 1145, DOI 10.1007/978-1-4939-0446-4_3, © Springer Science+Business Media New York 2014

agarose or polyacrylamide gel electrophoresis where dominant marker assays are scored based on the presence/absence of bands on a gel, while PCR products generated from codominant assays differ in the sizes of the bands due to length polymorphisms of the marker between the genotypes [3]. However, the major disadvantage of classical gel-based assays is that gel electrophoresis is unable to provide the high level of resolution required for distinguishing PCR products with very small length polymorphisms (<10 bp) or single nucleotide polymorphisms (SNP) without the addition of an intermediate step such as enzymatic cleavage of the PCR product at the site of the polymorphism. Given that not every SNP generates a restriction enzymatic cleavage site, classical gel-based assays may not always be applicable for assaying particularly troublesome SNPs.

The TSP technique was designed to overcome this problem and allows researchers to reliably distinguish SNP by gel electrophoresis after performing a simplified biphasic PCR protocol that utilizes two pairs of primers: the locus-specific (LS) primer pair and the nested locus-specific (NLS) primer pair [9–11]. The two primer pairs are separated by a large difference in annealing temperatures thus allowing both primer pairs to be present in the same PCR reaction mix but still only be involved in one phase of the PCR or the other, due to the temperature switch protocol. The initial PCR phase uses a high annealing temperature allowing the LS primer pair to bind to annealing sites that flank the region of interest on the template DNA, producing a PCR product larger than 400 bp in length and containing the SNP of interest. The second PCR phase uses a lower annealing temperature allowing the NLS primer pair to anneal to the target SNP for the detection of SNP allele-specific products. This chapter focuses on the workflow required for generating TSP markers, running TSP protocols and the technical aspect of using TSP markers to successfully achieve results that are easily separated by agarose gel electrophoresis.

2 Materials (*see* Note 1)

2.1 Tools and Instruments

1. Plastic tubes (10, 50 mL), plastic microfuge tubes (1.5, 2.0 mL).
2. Conical flasks and glass beakers.
3. Micropipettes.
4. Multichannel-micropipettes.
5. 96-Well plates (*see* **Note 2**) and adhesive seals.
6. 384-Well PCR plates and adhesive plate seals.
7. 384-Well PCR thermocycler.
8. Agarose gel casting tray and gel comb.
9. Agarose gel electrophoresis tank.
10. UV transilluminator equipped with a camera.

2.2 Reagents

1. Template DNA (20 ng/μL) (*see* **Note 3**).
2. Immolase™ DNA Polymerase (Bioline) (*see* **Note 4**), store at −20 °C.
3. Immolase™ PCR buffer with 1.5 mM $MgCl_2$ (Bioline) (*see* **Note 5**), store at −20 °C.
4. 50 mM $MgCl_2$ (Bioline) (*see* **Note 5**), store at −20 °C.
5. 1.25 mM dNTPs, store at −20 °C.
6. The forward and reverse locus-specific primers are mixed in a microfuge tube (1.5 mL) while the forward and reverse allele-specific (nested locus-specific) primers are to be mixed in another microfuge tube, store at −20 °C (*see* **Note 8**).
7. 10 mg/mL Bovine Serum Albumin, store at −20 °C.
8. Agarose, molecular grade (Bioline), store at room temperature.
9. SYBR® Safe gel stain (Invitrogen) (*see* **Note 6**), store at 2–25 °C.
10. 10× TAE (Tris-Acetate EDTA) buffer (Life Technologies): Weigh 48.4 g of Tris base (400 mM) (Tris (hydroxymethyl) aminomethane), 24 mL of absolute acetic acid (400 mM), and 3.7 g of EDTA (10 mM) (Ethylenediaminetetraacetic, disodium salt, dihydrate) and transfer to a 2 L glass beaker. Fully dissolve the Tris, acetic acid, and EDTA in 800 mL of Nanopure water then make up the solution to a final volume of 1 L with more Nanopure water (*see* **Note 7**). 1× TAE buffer is prepared from 10× TAE buffer by diluting 100 mL of 10× TAE with 900 mL of Nanopure water (*see* **Note 7**). Store both solutions at room temperature.
11. Loading buffer (Life Technologies): Dissolve 30 mg of bromophenol blue, 15.8 mg of Tris-HCl (1,3-Propanediol, 2-amino-2-(hydroxymethyl)-, hydrochloride), 37.2 mg of EDTA, and 6.5 g of sucrose in 10 mL of Nanopure water and vortex the solution until both components are fully dissolved and the solution is homogenous. Divide the solution into 1 mL aliquots and store at −20 °C.

3 Methods (*see* Note 1)

The TSP SNP assay utilizes two pairs of primers—the locus-specific (LS) primer set and the nested locus-specific (NLS) primer set—in a closed-tube biphasic PCR reaction to generate SNP allele-specific PCR products that can be distinguished by length polymorphisms using gel electrophoresis [11].

3.1 TSP Primer Design

In order to successfully obtain SNP allele-specific PCR products that can be differentiated by length polymorphisms, the LS and NLS primer pairs must be designed strictly to the following guidelines:

1. The LS primer pair must be designed based on DNA sequence flanking the target SNP to be interrogated. Parameters for LS primer pair design are:
   - Melting temperature of the LS primer pair must be between 60 and 65 °C
   - The PCR product size must be at least 400 bp
2. The NLS forward primer consists of two regions—a 5′ non-complementary tail region and a core complementary region. Parameters for NLS primer pair design are:
   - Melting temperature of the core complementary region of the NLS forward primer must be between 43 and 47 °C
   - The 5′ non-complementary tail region of the NLS forward primer should ideally increase the melting temperature of the primer to 53 °C
   - The 3′ terminal nucleotide of the NLS forward primer must be designed to anneal to the target SNP being interrogated
   - The melting temperature of NLS reverse primer should be between 52 and 55 °C but should optimally be 53 °C
   - The PCR product size must be less than 300 bp

   Primer design can be performed using Primer 3 version 4.0 (http://frodo.wi.mit.edu/) and NetPrimer software (http://www.premierbiosoft.com/netprimer/index.html) [13, 14].

3.2 TSP Assay

1. Prepare the 384-PCR plate and adhesive plate seals.
2. TSP components for one reaction: Each TSP reaction mixture contains 1.25 μL of 10× Immolase™ PCR buffer with 1.5 mM $MgCl_2$ (Bioline), 0.25 μL of 50 mM $MgCl_2$, 2 μL of 1.25 mM dNTPs, 0.125 μL of 10 mg/mL Bovine Serum Albumin, 0.125 μL of 10 μM LS forward/reverse locus-specific primer mix, 0.3125 μL of 20 μM NLS allele-specific primer mix, 0.05 μL of 5 U/μL Immolase™ DNA polymerase (Bioline), 3 μL of DNA template (20 ng/μL), and 5.385 μL of sterile Nanopure water. The total reaction volume should be 12.5 μL (*see* **Note 8**).
3. Apply the adhesive plate seal carefully ensuring that the plate is well-sealed (*see* **Note 9**).
4. Briefly vortex the PCR plate to mix the reactions well, then briefly centrifuge the PCR plate to ensure that the reaction mix settles to the bottom of the wells (*see* **Note 10**).

5. The thermal cycling protocol for TSP is as follows:

   - Denaturation at 95 °C for 3 min
   - Initial reaction phase: 15 cycles with each cycle at 95 °C for 30 s, 58 °C for 30 s, 72 °C for 60 s. This allows for the enrichment of the target locus harboring the expected SNP using the LS primer pair.
   - Second reaction phase: 5 cycles with each cycle at 95 °C for 10 s and 45 °C for 30 s; followed by 15 cycles with each cycle at 95 °C for 10 s, 53 °C for 30 s, and 72 °C for 5 s.

6. Upon completion of the thermal cycling protocol, briefly centrifuge the PCR plate (*see* **Note 10**). The PCR products can be stored on ice or at 4 °C until required for gel electrophoresis.

3.3 Agarose Gel Tank and Gel Comb Preparation

1. Select a gel casting tray and gel comb deep enough to cast an agarose gel that allows 8–10 μL of PCR product to be loaded per well. Ensure that the agarose gel to be casted will have sufficient lanes for the number of samples to be run as well as a few additional wells for running a DNA size marker ladder of choice.
2. Insert the gel comb into the gel casting tray and double-check that it is correctly positioned.

3.4 Agarose Gel Preparation (for a 24-Well Gel)

1. A 2 % agarose gel can be used to separate the TSP products as it is able to easily resolve the small DNA fragments (0.2–1 kb) of the TSP protocol. Weight 2 g of agarose and transfer it to a 500 mL conical flask.
2. Add 100 mL of 1× TAE and swirl to mix the reagents (*see* **Note 11**).
3. Dissolve the agarose by heating (generally performed by heating in a microwave). When completely dissolved, the agarose solution will be completely transparent (*see* **Note 12**).
4. Leave the agarose mixture to cool to about 55–60 °C (*see* **Note 13**).
5. Quickly add 10 μL of SYBR® Safe DNA gel stain to the agarose mixture and gently swirl to mix.
6. Pour the agarose mixture slowly into the gel casting tray to avoid bubbles. Any bubbles formed can be removed by using a disposable pipette tip.
7. Allow the agarose gel to set completely in the dark.
8. When the agarose gel has fully solidified, put the gel into the gel electrophoresis tank containing 1× TAE buffer. Ensure that the gel is completely immersed in the buffer (*see* **Note 7**).

3.5 Sample Loading

1. Add 2 μL of loading buffer into each PCR product sample that you wish to load onto the agarose gel and mix gently until the solution is homogeneous.
2. If required, load the first well on the agarose gel with a DNA size marker ladder of choice so that the sizes of the PCR products can be easily estimated (*see* **Note 14**).
3. Continue loading each individual sample into its respective well. If required, load the final well of the gel with the DNA size marker ladder of choice (*see* **Note 14**). When loading a large number of samples, one may choose to use a multichannel-micropipette to reduce the length of time required for sample loading as well as to prevent the loaded samples from diffusing out from the wells over a prolonged period of time.

3.6 Agarose Gel Electrophoresis

1. Run the gel at a constant 100 V. The length of the gel electrophoresis procedure is dependent on the size of the gel and percentage of agarose used. In general, the gel can be run for 45 min and the PCR products visualized on a UV/blue-light transilluminator (*see* **Notes 15** and **16**).
2. If the PCR products are not sufficiently resolved, the gel can be placed back into the agarose gel electrophoresis tank and ran at a constant 100 V for a longer period of time until sufficient resolution of the PCR products is achieved.
3. In general, the gel electrophoresis process can be stopped when the bromophenol blue front (from the loading buffer) has migrated approximately 2/3–4/5 the length of gel (*see* **Note 17**).
4. After visualizing the PCR products in a UV/blue-light transilluminator, one may wish to use the UV transilluminator camera to obtain a high-quality image of the agarose gel (in .tiff or .jpeg file formats) (*see* **Notes 15** and **16**).

3.7 TSP Product Detection

- The presence of the SNP allele interrogated is denoted by presence of the smaller PCR fragment on the gel that is amplified by the NLS primer pair, which is allele-specific and interrogates the target SNP allele (Fig. 1) (*see* **Note 18**).
- The presence of the alternate SNP allele is denoted by the presence of the larger PCR fragment on gel that is amplified by the LS forward primer and NLS reverse primer. The presence of this larger PCR fragment not only denotes the absence of the target SNP allele but can also be considered a "positive control" when compared to a failed PCR assay (Fig. 1).
- In heterozygote samples, both the target SNP and its alternate allele will be detected and this is denoted by the presence of both the smaller allele-specific PCR fragment and the larger alternate SNP allele PCR fragment on the agarose gel (Fig. 1).

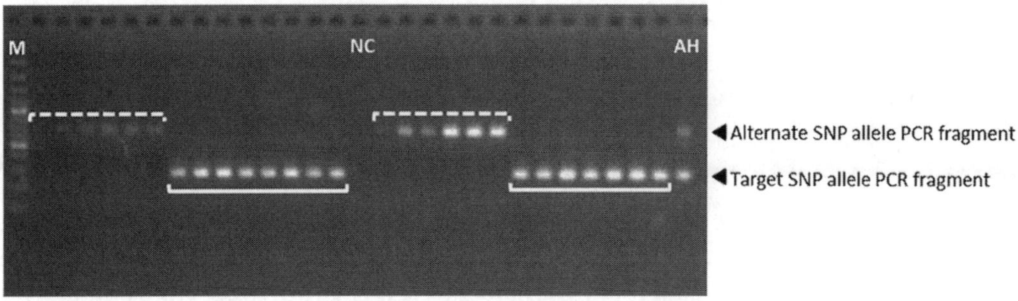

Fig. 1 Agarose gel electrophoresis of PCR products obtained from a TSP marker assay screened across 27 Australian barley varieties. The presence of the targeted SNP allele is denoted by the presence of the smaller PCR fragment (*highlighted* with *solid upturned brackets*), while the presence of the alternate SNP allele is denoted by the presence of the larger PCR fragment (*highlighted* by the *dashed downturned brackets*). In the case of a heterozygote, both PCR fragments are present. M represents the Fermentas GeneRuler™ 100 bp size ladder; NC represents a no-template DNA negative control; AH represents an artificial heterozygote made using a 1:1 ratio of genomic DNA from an Australian variety known to carry the target SNP allele and another variety known to carry the alternate SNP allele

4 Notes

1. For occupational health, safety, and welfare reasons, laboratory gloves and a lab coat should be worn at all times when performing experiments. Other personal protective equipment such as full face-shields may also be necessary for certain procedures such as agarose gel preparation.
2. To shorten experimental setup time and operator error, one may choose to prepare the master mix and DNA samples in 96-well plates so that multichannel-micropipettes or a programmed liquid-handling robot can be used to dispense the master mix/samples into the 384-well PCR plates.
3. The protocol must be conducted with a small amount of high-quality DNA. The phenol–chloroform DNA extraction protocol from [12] is recommended as it is able to extract substantial amounts of high quality DNA. The DNA pellet can be dissolved in sterile Nanopure water and store at −20 °C (or −80 °C for longer term storage). Quantification of DNA can be performed with a NanoDrop spectrophotometer and adjusted to a final concentration of 20 ng/μL.
4. To ensure that the enzyme remains active and stable even after prolonged storage, it should always be kept at −20 °C and should be added last to the TSP PCR reaction. Other *Taq* DNA polymerases such as *Tfi* DNA polymerase [11] can be used to substitute Immolase™ DNA polymerase.
5. The 50 mM $MgCl_2$ solution is used to achieve a final concentration of 2.5 mM $MgCl_2$ in the final TSP PCR reaction mix.

Increasing the $MgCl_2$ concentration can substantially improve the enzymatic activity of the Immolase™ DNA polymerase.

6. Although ethidium bromide can also be used to visualize double-stranded DNA PCR products in gels, the authors recommend using SYBR® Safe as a less hazardous chemical replacement for ethidium bromide. However, SYBR® Safe is light-sensitive thus should be exposed to light as little as possible. Samples containing SYBR® Safe should be wrapped in aluminum foil and gel tanks used for gel electrophoresis of PCR products to be visualized with SYBR® Safe should be shielded from light sources as much as possible.
7. For convenience, one may choose to make a stock of 10× TAE buffer for future use. Undissolved materials seen during the preparation of the 10× TAE buffer solution could be dissolved by heating the solution in a hot water bath (approximately 60–70 °C). 1× TAE solution used for gel electrophoresis should be replaced every 2 runs to ensure that the DNA bands migrate properly through the agarose gel and to prevent the accumulation of waste products from previous runs.
8. In order to account for pipetting and operator errors, the master mix (containing all the reagents minus the template DNA) can be prepared at a slightly larger volume, normally 5–10 % more than the total volume of the number of reactions required for the experiment. The master mix should be assembled in the following order: Immolase™ PCR buffer, $MgCl_2$, dNTP, Bovine Serum Albumin, LS and NLS primer mixes, Immolase™ DNA polymerase, and made up to the final volume with sterile Nanopure water. Optimal PCR results are obtained when the concentration ratio of NLS to LS is about 5:1 [11].
9. Ensure that the adhesive seal covers the 384-well PCR plate then apply pressure with a hard flat surface around the skirting of the PCR plate to ensure that the plate is properly sealed. Pressure should also be applied to the top of the wells to ensure that a seal is formed around each well. Proper sealing of the plate is essential for reducing evaporation and preventing the seal from popping during the PCR thus causing cross contamination of the PCR products.
10. Vortexing of the samples should only be done gently and briefly to prevent shearing of the genomic DNA and Taq polymerase enzyme molecules. Centrifugation of the PCR plate ensures that the samples are collected at the bottom of the wells and minimizes sample volume loss.
11. Mixing of the reagents prior to heating the mixture avoids coagulation and sedimentation of the agarose at the bottom of the flask which will then take longer to dissolve into solution.

12. Microwave-heating of the mixture may cause superheating of the mixture. A full face-shield, safety glasses, gloves and a lab coat should be worn when handling the heated mixture.
13. The gel mixture should be swirled gently every 1–2 min during the cooling process to ensure uniform cooling throughout the mixture.
14. The choice of which DNA size marker ladder to use is determined by the expected sizes of the PCR products. Given that the expected product sizes of TSP marker assays are between 200 and 500 bp, a DNA size marker ladder with higher resolution at this molecular weight range, such as Bioline HyperLadder™ II, is preferable.
15. Avoid exposing the agarose gel to UV light for more than 3–5 min as the intensity of the SYBR® Safe stain quickly diminishes in the presence of UV light. The Safe Imager™ blue-light transilluminator is recommended for capturing gel images as it does not diminish the intensity of the SYBR® Safe signal as quickly as UV light. It is also a safer alternative to UV light.
16. Wear personal protective equipment such as a full face-shield and a lab coat to limit exposure to UV light.
17. The bromophenol blue molecules carry a negative charge and will migrate in the same direction as the DNA molecules in the agarose gel. Generally speaking, the bromophenol blue front should migrate at the same rate as a DNA fragment of approximately 150 bp in a 2 % agarose gel ran in 1× TAE solution.
18. In order to ensure that the results of the TSP protocol are accurate and replicable, replicates of each DNA samples as well as positive and negative controls should be included into the experimental design.

References

1. Rostoks N, Mudie S, Cardle L et al (2005) Genome-wide SNP discovery and linkage analysis in barley based on genes responsive to abiotic stress. Mol Genet Genomics 274:515–527
2. Collard BCY, Jahufer MZZ, Brouwer JB et al (2005) An introduction to markers, quantitative trait loci (QTL) mapping and marker-assisted selection for crop improvement: the basic concepts. Euphytica 142:169–196
3. Semagn K, Bjørnstad Å, Ndjiondjop MN (2006) An overview of molecular marker methods for plants. Afr J of Biotechnol 5:2540–2568
4. Acquaah G (2007) Principles of plant genetics and breeding. Blackwell Publishing, UK
5. Von Korff M, Grando S, Del Greco A et al (2008) Quantitative trait loci associated with adaptation to Mediterranean dryland conditions in barley. Theor Appl Genet 117:653–669
6. Collins NC, Tardieu F, Tuberosa R (2008) Quantitative trait loci and crop performance under abiotic stress: where do we stand? Plant Physiol 147:469–486
7. Farooq M, Wahid A, Lee DJ et al (2009) Advances in drought resistance of rice. Crit Rev Plant Sci 28:199–217
8. Fleury D, Jefferies S, Kuchel H et al (2010) Genetic and genomic tools to improve drought tolerance in wheat. J Exp Bot 61:3211–3222
9. Hayden MJ, Tabone T, Mather DE (2009) Development and assessment of simple PCR markers for SNP genotyping in barley. Theor Appl Genet 119:939–951

10. Hayden MJ, Tabone TL, Nguyen TM et al (2010) An informative set of SNP markers for molecular characterisation of Australian barley germplasm. Crop Pasture Sci 61:70–83
11. Tabone T, Mather DE, Hayden MJ (2009) Temperature Switch PCR (TSP): Robust assay design for reliable amplification and genotyping of SNPs. BMC Genomics. doi:10.1186/1471-2164-10-580
12. Rogowsky PM, Guidet FLY, Langridge P et al (1991) Isolation and characterization of wheat-rye recombinants involving chromosome arm 1DS of wheat. Theor Appl Genet 82:537–544
13. Rozen S, Skaletsky HJ (2000) Primer3 on the www for general users and for biologist programmers. In: Krawet S, Misener S (eds) Bioinformatics methods and protocols: methods in molecular biology. Humana Press, Totowa, pp 365–386
14. NetPrimer [http://www.premierbiosoft.com/netprimer/netprlaunch/netprlaunch.html]
15. De Koeyer D, Douglass K, Murphy A et al (2010) Application of high-resolution DNA melting for genotyping and variant scanning of diploid and autotetraploid potato. Mol Breeding 25:67–90

Chapter 4

Multiplex-Ready Technology for Mid-Throughput Genotyping of Molecular Markers

Julien Bonneau and Matthew Hayden

Abstract

Screening molecular markers across large populations in breeding programs is generally time consuming and expensive. The Multiplex-Ready Technology (MRT) (Hayden et al., BMC genomics 9:80, 2008) was created to optimize polymorphism screening and genotyping using standardized PCR reaction conditions. The flexibility of this method maximizes the number of markers (up to 24 markers SSR or SNP, ideally small PCR product <500 bp and highly polymorphic) by using fluorescent dye (VIC, FAM, NED, and PET) and a semiautomated DNA fragment analyzer (ABI3730) capillary electrophoresis for large numbers of DNA samples (96 or 384 samples).

Key words SSR, SNP, Multiplexing, Capillary electrophoresis

1 Introduction

Marker assisted selection uses genetic information at known loci to track desirable alleles in plant breeding programs. Multiplexing methods enable several loci to be tracked at the same time, which is particularly convenient for gene-pyramiding. Multiplex PCR, in which more than one marker is simultaneously amplified in the same reaction, increases the amount of information generated per assay to reduce consumable and labor costs. However, the optimization of a multiplex PCR can be time-consuming and complicated due to different annealing temperature for each primer pair and requirement for different primer concentrations. The multiplex PCR can also encounter other problems such as undesirable primer–primer interactions and nonspecific amplification. The M13-tailed primer method [1] is commonly used for the assay of SSRs. This three primer strategy consist of performing a PCR using a forward primer with a nucleotide extension at its 5′-end. This extension is complementary to a fluorescently labelled M13 reverse primer where the fluorescence is activated once the first few cycles

Delphine Fleury and Ryan Whitford (eds.), *Crop Breeding: Methods and Protocols*, Methods in Molecular Biology, vol. 1145, DOI 10.1007/978-1-4939-0446-4_4, © Springer Science+Business Media New York 2014

are performed; however the M13-tailed primer can often generate nonspecific products.

Based on standardized polymerase chain reaction (PCR) using capillary electrophoresis coupled with fluorescence-based detection, MRT [2] facilitates the screening of markers (up to 24) across large numbers of germplasm. SSR and SNP markers are best suited for this application because of the small size of their PCR product (<500 bp) and high level of polymorphism. The method described here combines the advantages of the M13-tailed primer method and the property of multiplexing PCR products using fluorescence. The method requires the synthesis of primers in the multiplex-ready format for its use. This technology has been successfully used in wheat [3], *Panicum virgatum* [4], *Medicago truncatula* [5], *and Lupinus angustifolius* [6].

2 Materials

All PCR reagents should be kept in ice. Fluorescent reagents should be protected from light (e.g., wrapped in aluminum foil), as they are rapidly photo-bleached. Prepare and store all reagents at −20 °C unless indicated otherwise. Follow waste disposal regulations when disposing of waste materials.

2.1 Polymerase Chain Reaction (PCR) Components

1. Preparation of 100× BSA (Bovine Serum Albumin).
   Table 1 describes how to prepare 100× BSA. Once components have dissolved, make up to 50 ml final volume with sterile water and pass through a 50 μm filter. Store 1 ml aliquots at −20 °C in a constant-temperature freezer.
2. Preparation of 5× Multiplex-Ready Buffer (5× Mpx-Rdy Buffer). Table 2 describes how to prepare 40 ml of 5× Multiplex-Ready PCR buffer. Immolase buffer (1× buffer contains 16 mM $(NH_4)_2SO_4$, 0.01 % Tween-20, 100 mM Tris–HCl, pH 8.3) is usually provided with Immolase enzyme from Bioline or equivalent. To ensure high assay reproducibility, it is recom-

Table 1
Preparation of 100× BSA

		Final concentration
0.500 g	Bovine serum albumin fraction V	10 mg/ml
0.145 g	Sodium chloride	50 mM
10 μl	0.5 M EDTA, pH 8.0	0.1 mM
2.5 ml	Glycerol	5 %
30 ml	Sterile water	

Table 2
5× Multiplex-Ready PCR buffer

		Final concentration
20 ml	10× Immolase PCR buffer (Bioline)	5×
6 ml	50 mM $MgCl_2$ (Bioline)	7.5 mM
0.4 ml	100 mM dATP (Bioline)	1 mM
0.4 ml	100 mM dCTP (Bioline)	1 mM
0.4 ml	100 mM dTTP (Bioline)	1 mM
0.4 ml	100 mM dGTP (Bioline)	1 mM
1 ml	100× BSA	2.5×
11.4 ml	Sterile water	
40 ml	Total	

mended that the multiplex-ready PCR buffer is prepared in bulk and stored as 1 ml aliquots at −20 °C in a constant-temperature (not frost-free) freezer.

3. Preparation of ultrafiltration (UF) plate.
   Pre-wet the ultrafiltration plate (AcroPrep® 384 Filter Plate with 10 kDa Omega membrane, PALL Life Sciences) by adding 20 μl of sterile water to each well. Gently tap the plate on the bench to dislodge any air bubbles on the membrane surface. Let the plate stand for 5 min at room temperature then remove the excess water by tapping the inverted plate on paper towel (*see* **Note 1**).
4. Preparation of formamide for ABI3730.
   Add 800 μl of HiDi™ Formamide (Applied Biosystems®) and 4 μl of GeneScan™ 500LIZ™ size standard (Applied Biosystems®) to a 1.5 ml tube. This volume suits the assay of 96 samples.
5. The master mix for marker optimization contains all PCR reagents, except locus-specific primers (Table 3; *see* **Note 2**). DNA samples should be diluted in buffers without EDTA, such as 10 mM Tris–HCl pH8.0. EDTA chelates magnesium ions and may result in suboptimal marker amplification.
6. Master mix for genetic mapping.
   This master mix contains all PCR reagents, except the DNA. DNA samples should be added separately for each sample and diluted in buffers without EDTA, such as 10 mM Tris–HCl pH 8.0. EDTA chelates magnesium ions and may result in suboptimal marker amplification. Table 4 describes how to prepare a master mix (MM2) for marker screening and genetic mapping.

Table 3
Preparation of the master mix for selection of the optimal locus-specific primer concentration

1× Master mix 1 (MM1)	Final concentration	Volume required (μl)
5× Mpx-Rdy Buffer	1×	1.2
Immolase (5 U/μl)	0.15 U	0.03
10 μM dye-labelled *tagF* primer	75 nM	0.041
10 μM *tagR* primer	75 nM	0.045
Genomic DNA (50 ng/μl)	50 ng	1
Sterile water		0.68
Locus-specific concentration	Variable	3
Total		6

Table 4
Preparation of the master mix for one locus-specific primer according to its optimal concentration

1× master mix 2 (MM2)	Final concentration	Volume required (μl)
5× Mpx-Rdy Buffer	1×	1.2
Immolase (5 U/μl)	0.15 U	0.03
10 μM dye-labelled *tagF* primer	75 nM	0.045
10 μM *tagR* primer	75 nM	0.045
4 μM locus-specific concentration	Variable	X
Sterile water		2.68–X
Total		4

2.2 Equipment

1. PCR machines (384- and 96-well format).
2. 384- and 96-well plates.
3. 1.5 ml microtubes.
4. 8 and 12 multichannel pipettes P20 and P200.
5. Vortex (tube and plate).
6. Centrifuge machine (Tube and plate).
7. Tips 10, 20, 200, and 1,000 μl.
8. AcroPrep® 384 Filter Plate with 10 kDa Omega membrane.
9. Vacuum manifold (e.g., MILLIPORE®).
10. Automated DNA fragment analyzer (e.g., ABI PRIM® 3730xl, Applied Biosystems).

11. GeneMapper® software (Applied Biosystems®).
12. Aluminum foil.
13. Laboratory cleansing paper.
14. PCR plate covers.

3 Methods

3.1 MRT Marker Optimization

The optimal concentration of locus-specific primer for MRT must be determined empirically for each new marker. Once determined, the optimal concentration is used irrespective of whether uniplex or multiplex assays are performed. The optimal concentration can be determined by assessing the yield of target PCR product and overall PCR specificity for multiplex-ready PCR assays performed with different concentrations of locus-specific primer. The optimal concentration is the lowest concentration producing strong amplification (*see* **Note 3**) of the target PCR product with the least amount of nonspecific amplification.

1. Dilute each locus-specific primer to 400 nM using 10 mM Tris–HCl, pH 8.0. Prepare 20 μl dilutions of 20, 40, 60, 80, 100, 120, 140, and 160 nM locus-specific primer (*see* **Note 4**).
2. Prepare a PCR master mix (Table 3) for each of the DNA tested.
3. Add 3 μl of master mix to the PCR plate, followed by 3 μl of locus-specific primer. Note that the final concentration of locus-specific in the multiplex-ready PCR will be halved.
4. Perform PCR as follows: 2 min initial denaturation at 94 °C to activate the DNA polymerase, followed by a total of 65 cycles: 30 s at 92 °C, 90 s at 50 °C, 60 s at 72 °C for five cycles, then 30 s at 92 °C, 90 s at 63 °C, and 60 s at 72 °C for 20 cycles, followed by 40 cycles with 15 s at 92 °C, 30 s at 54 °C, and 60 s at 72 °C, and a final extension step of 10 min at 72 °C.
5. Spin down the plate after performing the PCR. Add 12 μl of sterile water in each of the PCR wells, vortex and thoroughly pulse spin the samples (*see* **Note 5**).
6. Pool 6 μl (VIC), 6 μl (FAM), 3 μl (NED), and 6 μl (PET) μl of each PCR products for each primer tested directly into the UF plate (*see* **Note 6**).
7. Place the unsealed UF plate onto a vacuum manifold at least 21 inches of Hg for 5 min. (21 Hg corresponds to 700–750 mbar on the vacuum manifold pressure gauge). Release the pressure, dry the bottom of the plate with laboratory cleansing wipes, seal the plate, and repeat the step with the same conditions (*see* **Note 7**).

8. Add 18 μl of sterile water to dislodge the PCR product from the membrane of the UF plate, once the vacuum step is complete; leave the plate for 30 min on the bench at room temperature to allow the PCR products to resuspend.
9. Retrieve the PCR product into a new plate (96-well plate) by pipetting up and down 20 times to ensure the dissolution of PCR products (*see* **Note 8**).
10. Add 3 μl of desalted PCR product into a new plate (96-well plate compatible with ABI3730) with 8 μl of Formamide for ABI3730 in each well. Seal the plate (*see* **Note 9**) and denaturate at 95 °C for 5 min using a 96-well PCR machine.
11. Place the plate onto the DNA fragment analyzer (ABI3730) (*see* **Note 10**).
12. Read the output file using GeneMapper™ to deconvolute the markers and the colors (*see* **Note 11**).

3.2 Determining Marker Allele Size Range

After identifying the optimal concentration per primer pair, you need to characterize the range of allele sizes per marker. The marker should be tested on 6–8 genetically diverse lines to identify the smallest and the largest amplicon (PCR product bp) to set an allele size range per marker. This will enable pooling of markers with different size range and distinguishing of those of with similar size by using the four different dyes. The MRT optimum is 24 markers per DNA sample.

1. Prepare 1 ml of primer dilution to 400 nM.
2. Prepare master mix (Table 3) for the number of genetically different lines tested (6–8 DNA samples).
3. Proceed from **steps 3–12** as above (Subheading 3.1).
4. Identify the allele size range for each of the primers tested using GeneMapper™ software (*see* **Note 11**).

3.3 Running MRT by Capillary Electrophoresis

Cost efficient genotyping on a semiautomated DNA fragment analyzer requires that several markers can be co-analyzed. This can be achieved by pooling single PCR amplifications. The only requirement is that PCR fragment sizes of markers within the same dye detection channel do not have overlap. The use of the BINNER software (http://www.genica.net.au/) to construct multiplex-ready marker panels (Table 5; *see* **Note 12**) will ensure that this criterion is met (for further information about the BINNER software see http://www.genica.net.au/).

The example described here fulfils the maximum that can be performed on 96 DNA samples using 24 markers. *See* Fig. 2 which shows the method step by step. The four labelled primers are represented by color code (Fig. 1b, c): VIC, FAM, NED, and PET from left (light grey) to right (dark grey). The tubes and plates

Table 5
Description of the four panels that include six markers for each of the following dye: FAM, VIC, PET, and NED in GeneMapper software screenshot

Panel	Dye	Left size limit (bp)	Right size limit (bp)
Marker 1	FAM	107	122
Marker 2	FAM	127	187
Marker 3	FAM	191	251
Marker 4	FAM	287	307
Marker5	FAM	310	320
Marker6	FAM	260	280
Marker7	VIC	97	111
Marker8	VIC	114	196
Marker9	VIC	202	223
Marker10	VIC	240	245
Marker11	VIC	260	297
Marker12	VIC	306	330
Marker13	NED	102	117
Marker14	NED	121	194
Marker15	NED	262	280
Marker16	NED	286	298
Marker17	NED	196	254
Marker18	NED	301	322
Marker19	PET	100	120
Marker20	PET	122	193
Marker21	PET	310	330
Marker22	PET	287	300
Marker23	PET	200	256
Marker24	PET	263	283

The range of the left size limit and right size limit indicate the range of the expected PCR product that should be visualized (Example in Fig. 2)

containing fluorescent dye must be protected against light using aluminum foil.

1. Distribute four times the 96 DNA samples in each quarter of the six plates (384-well format) (Fig. 1b).

a

b

c

d

e

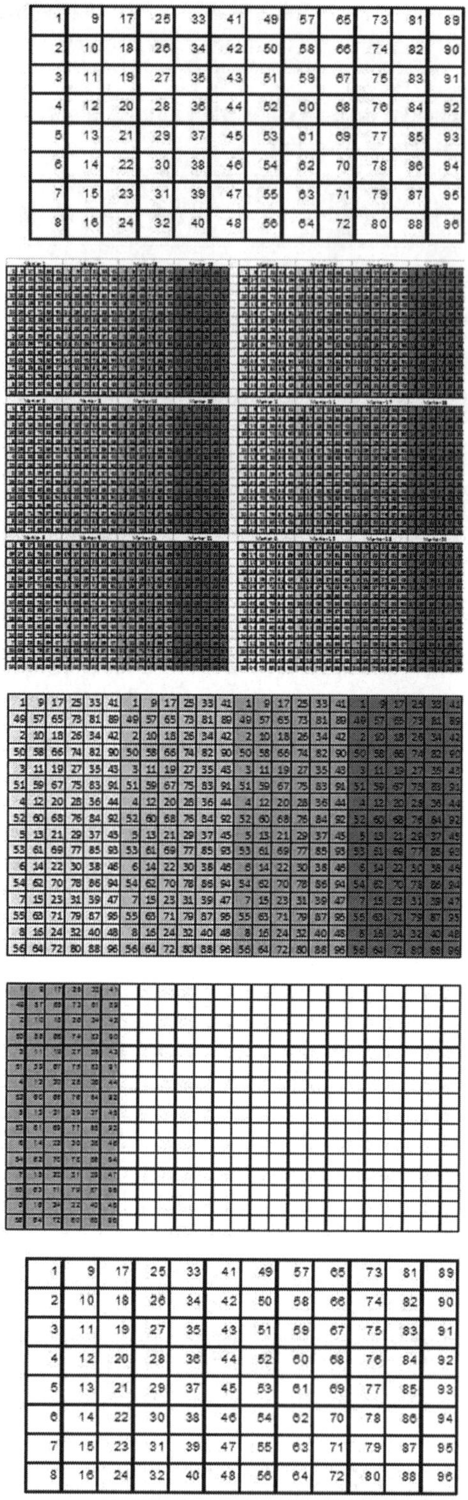

Fig. 1 Suggestion of the methods through the methods: (**a**) plate of the 96 DNA samples organization to be tested, (**b**) the six 384-well plate format where the DNA is distributed according to the grey-scale color code for each dye VIC, FAM, NED, and PET. (**c**) The six plate quarters containing markers with the same dye are pooled in the pooling plate, where each quarter contains six markers per color, (**d**) The pooling plate is then re-pooled in the ultrafiltration plate which containing the following ratio of each dye 2:2:1:2 for VIC (*green*), FAM (*blue*), NED (*red*), and PET (*yellow*), and (**e**) the storage plate where the PCR can be stored at –20 °C allowing rerun if needed

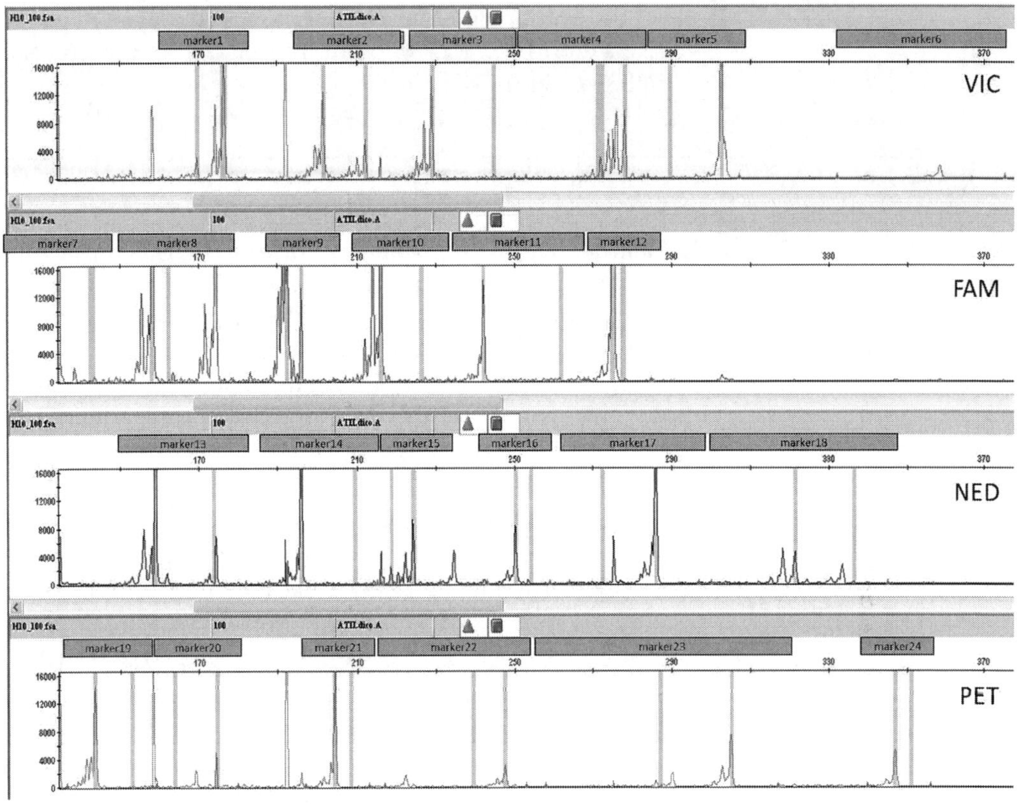

Fig. 2 Visualization of the output of the capillary electrophoresis (ABI3730) for one DNA sample run with 24 markers and 4 dyes

2. Make up a dilution of 4 μM for each of the 24 markers (*see* **Note 12**; Table 5).
3. Prepare 24 master mixes (*see* Tables 4 and 5) with the four labelled primers (VIC, FAM, NED, and PET) (*see* **Note 13**).
4. Run the PCR program as described in Subheading 3.1.
5. Spin down the plate after performing the PCR. Add 12 μl of sterile water in each of the PCR wells, vortex and thoroughly pulse spin the samples (*see* **Note 5**).
6. Retrieve 4 μl of each PCR product that contains the same dye into an empty 384-well plate, each of the quarter will contain 24 μl of PCR products from 6 different markers (Fig. 2c; *see* **Note 6**).
7. Pool 6, 6, 3, and 6 μl of each of the pooled PCR product for each dye directly into the UF plate (*see* **Note 6**; Fig. 1d).
8. **Steps 8–10** as described Subheading 3.1.
9. **Step 11** as described in Subheading 3.1 and as shown in Fig. 1d.
10. **Step 12** as described in Subheading 3.1.

11. Read the output file using GeneMapper™ to deconvolute the markers and the colors to score markers across the 96 DNA samples (*see* **Note 11**; Fig. 2).

4 Notes

1. Ultrafiltration of PCR products prior to separation on an automated DNA fragment analyzer is required to remove reaction salts, unincorporated dNTPs and PCR products less than 100-bp in size. Salt ions strongly impede the uptake of PCR products on most automated DNA fragment analyzers, particularly those employing electrokinetic injection, and significantly reduce genotyping quality.
2. The 400 nM locus-specific primer stocks are prepared in advance for each marker. The marker must be added to the multiplex-ready reaction at two times their required final concentration. This higher primer concentration is required to account for the twofold primer dilution that is achieved when 3 μl of diluted locus-specific primer is added to 3 μl of PCR master mix.
3. Using automated DNA fragment analyzer (e.g., ABI PRIM® 3730xl, Applied Biosystems), the peak height should be relatively uniform (e.g., 3,000–6,000 RFU range florescent unit). This reference might vary due to the use of different platforms. However, as long as the peak heights are constant across markers the remaining steps can be proceed.
4. The locus-specific primer solution comprises an equimolar amount of the forward and reverse primers designed for the target locus. The forward and reverse primers need to be synthesized with the nucleotide sequence 5′ ACGACGTTGTAAAA 3′ and 5′ CATTAAGTTCCCATTA 3′, respectively, at their 5′ end. In addition, two generic tag primers, tagF and tagR, with sequences 5′ ACGACGTTGTAAAA 3′ and 5′ CATTAAGT TCCCATTA 3′, respectively, need to be synthesized with the tagF primer labelled at its 5′-end with one of the following fluorescent dyes: VIC, FAM, NED, PET, and HEX (available from Applied Biosystems). Ideal markers are single locus (amplifying only one PCR product); however it is possible to use multi-locus primers (more than one PCR product that belongs to different physical location in the genome). Based on the PCR product size, markers can be co-separated on the capillary gel as long as their PCR product sizes don't overlap. If the PCR products overlap, they can then be easily separated by using different dyes. MapManager Software (Applied Biosystems) is able to de-convolute the output and separate

PCR products on the basis of fluorescent dye colour and size (Table 4).

5. A 12 multichannel pipette can be used to facilitate the water distribution.
6. There is no need to change the tips of the multichannel pipette since the DNA is the same between markers of the same dye color. The pooling plate combines the four dyes at a ratio of 2:2:1:2, respectively, to compensate the relative fluorescence of each dye when detected during electrophoresis.
7. This step is supposed to ensure the total evacuation of liquid. However, it is possible that some PCR product still remain in the plate. The retention volume should be around 1.5 μl, if, the volume is higher than 1.5 μl then repeat the step.
8. At this stage the plate can be stored at −20 °C.
9. The seal is essential to avoid any evaporation.
10. The plate can be kept in the fridge or freezer until reading with light protector to avoid degradation.
11. The allele scoring requires experience and a training session of the GeneMapper software is recommended and available at http://www.genica.net.au/.
12. In the case of less than six markers per color it is recommended to use "dummy" reaction, by adding water in empty wells.
13. It is recommended to prepare 1 ml of 400 nM equimolar solution of locus-specific primers which can be kept at −20 °C for future experiments.

Acknowledgment

We thank Thao M. Nguyen, Amanda Waterman, and Kenneth J. Chalmers (University of Adelaide).

References

1. Oetting WS, Lee HK, Flanders DJ et al (1995) Linkage analysis with multiplexed short tandem repeat polymorphisms using infrared fluorescence and M13 tailed primers. Genomics 30: 450–458
2. Hayden MJ, Nguyen TM, Waterman A et al (2008) Multiplex-ready PCR: a new method for multiplexed SSR and SNP genotyping. BMC Genomics 9:80
3. Bonneau J, Taylor J, Parent B et al (2013) Multi-environment analysis and improved mapping of a yield-related QTL on chromosome 3B of wheat. Theor Appl Genet 126:747–761
4. Liu LL, Wu YQ (2012) Development of a genome-wide multiple duplex-SSR protocol and its applications for the identification of selfed progeny in switchgrass. BMC Genomics 13:522
5. Kamphuis L, Gao L, Singh K (2012) Identification and characterization of resistance to cowpea aphid (Aphis craccivora Koch) in Medicago truncatula. BMC Plant Biol 12:101
6. Gao LL, Hane JK, Kamphuis LG et al (2011) Development of genomic resources for the narrow-leafed lupin (*Lupinus angustifolius*): construction of a bacterial artificial chromosome (BAC) library and BAC-end sequencing. BMC Genomics 12:521

Chapter 5

Genotyping by High-Resolution Melting Analysis

Elise J. Tucker and Bao Lam Huynh

Abstract

High-resolution melting (HRM) analysis is a simple, closed tube, post-PCR method used to identify genetic variation. The method is highly sensitive and can discriminate DNA sequence variants based on length (such as insertions or deletions), composition (such as single nucleotide polymorphisms, i.e., SNP) or strand complementarity (such as heterozygous or homozygous material). The technique involves PCR amplification of a target sequence in the presence of a fluorescent double-stranded DNA (dsDNA) binding dye, melting of the fluorescent amplicons, and subsequent interpretation of melt curve profiles. Here, we describe general considerations for assay design, PCR amplification, and HRM analysis.

Key words High-resolution melting curve analysis, HRM, Single nucleotide polymorphism, SNP genotyping, SYTO®9

1 Introduction

The technique of high-resolution melting (HRM) was first described in 2003 [1] and since then its use has become widespread due to its simplicity, sensitivity, cost effectiveness, and flexible design. Examples have been reported in diverse species including diploids [2, 3] and polyploids [4, 5]. Using the same experimental principles and techniques, HRM can be used for discovery of genetic variants, in so-called "gene scanning" experiments, or to genotype known sequence variants in optimized assays [1, 6]. HRM is suitable to detect polymorphisms such as SNP [7], length polymorphisms such as microsatellites [3], and insertion/deletions such as insertion site-based polymorphisms (ISBP) markers [5, 8]. HRM experiments involve three main steps: (1) PCR amplification of a target region to high copy numbers in the presence of a fluorescent double-stranded (ds) DNA binding dye, (2) melting of the fluorescently labelled product, and (3) analysis of melt curves.

The success of HRM as a technique has largely been due to the development of third generation dsDNA binding dyes and high

Delphine Fleury and Ryan Whitford (eds.), *Crop Breeding: Methods and Protocols*, Methods in Molecular Biology, vol. 1145, DOI 10.1007/978-1-4939-0446-4_5, © Springer Science+Business Media New York 2014

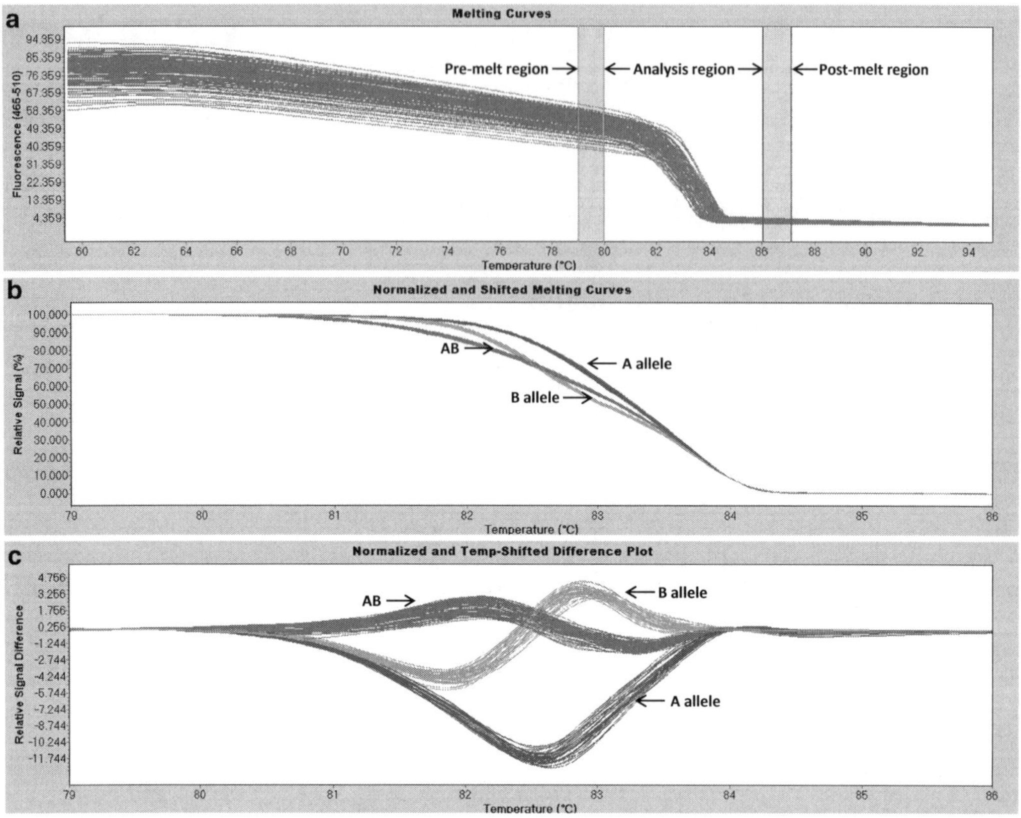

Fig. 1 Genotyping by high-resolution melting analysis. (**a**) Raw fluorescent melting curves HRM with pre-melt and post-melt sliders defining the region for analysis. (**b**) Normalized and temperature shifted melting curves. (**c**) Normalized and temperature shifted difference plots

precision melting instruments that are capable of acquiring large amounts of fluorescent data over incremental temperature changes [6]. Third generation dsDNA binding dyes, such as SYTO®9 (Life Technologies), EvaGreen™ dye (Biotium, Inc.), or LC Green® (BioFire Diagnostics, Inc.), can be used at higher concentrations to completely saturate dsDNA samples without inhibiting PCR, unlike second generation dsDNA intercalating dyes such as SYBR® Green I (Life Technologies) [9]. The dsDNA binding dye intercalates with the double-stranded PCR product and fluoresces brightly in this state at low temperatures. Following PCR, the fluorescently labelled dsDNA sample is melted through a range of temperatures on a specialized melting instrument, e.g., LightCycler® 480 Real Time PCR System (Roche), LightScanner® System (BioFire Diagnostics, Inc.) and fluorescence of the sample is measured at a minimum of 25 data acquisitions per degree change. As the DNA progressively dissociates into single-stranded DNA, fluorescence of the sample decreases until the sample is completely single-stranded and no fluorescence is detected. Analysis of the fluorescent data in instrument-specific software generates melt profiles that are characteristic and reflect the DNA composition of a given sample. Data can be visualized as raw melt curves, as normalized melt curves or as difference plots (Fig. 1).

Robust assays are paramount to the power, usefulness, and robustness of HRM for mutation discovery and genotyping. Primer design, reaction components, and cycling conditions are critical components of HRM success. Primers should amplify a specific product 100–300 bp in length, have a melting temperature (*T*m) of 60 °C, and produce no secondary structures. PCR reaction components and cycling conditions should be optimized to produce a specific high yielding product. HRM optimization strategies often involve testing multiple primer sets to identify the most robust pair, preparing uniform DNA template, titrating $MgCl_2$ concentration, titrating primer concentration, and altering PCR cycling conditions.

2 Materials

Below we describe PCR master mixes made from individual reaction components. It is, however, possible to buy ready-made master mixes (*see* **Note 1**). Prepare all solutions in sterile plasticware using sterile ultrapure water. Store all reagents at −20 °C. Follow local and institutional waste disposal regulations when disposing of waste materials.

2.1 PCR Components

1. DNA template (*see* **Note 2**).
2. Hot-start DNA polymerase (*see* **Note 3**): 5 U/μl IMMOLASE™ DNA polymerase (Bioline, catalogue number BIO-21048).
3. PCR buffer: 10× ImmoBuffer (Bioline) containing 160 mM $(NH_4)_2$ SO_4, 1 M Tris–HCl pH 8.3 and enhancers. Store at −20 °C.
4. Primers (10 μM): Combine 10 μl of 100 μM forward primer, 10 μl of 100 μM reverse primer and 80 μl of molecular biology grade water to make a 10 μM primer solution containing both forward and reverse primers.
5. 2 mM dNTP: 2 mM dATP, 2 mM dTTP, 2 mM dCTP, 2 mM dGTP in molecular biology grade water.
6. 50 mM $MgCl_2$.
7. Bovine serum albumin (BSA): 10 mg/ml (New England Biolabs, Inc., catalogue number B9001S).
8. dsDNA binding dye (50 μM): Dilute 10 μl of 5 mM SYTO®9 green fluorescent nucleic acid stain (Life Technologies, catalogue number S-34854) in 990 μl of molecular biology grade water. Make 200 μl aliquots and protect from light (*see* **Notes 4–7**).
9. Molecular biology grade sterile water (*see* **Note 8**).

10. PCR vessel: Use white coloured PCR vessels compatible with the available PCR and HRM instruments (*see* **Note 9**).
11. Seals: optically clear sealing film (*see* **Note 10**).

2.2 PCR Cycling and HRM

1. PCR thermal cycler.
2. Instrument capable of HRM analysis, e.g., LightCycler® 480 Real Time PCR System (Roche), LightScanner® System (BioFire Diagnostics, Inc.).

3 Methods

3.1 Primer Design

1. Use specialised software (*see* **Note 11**) to design primers (*see* **Notes 12** and **13**) that amplify a product 100–300 bp in length (*see* **Note 14**), have a melting temperature (Tm) of ~60 °C, a GC content ~40–60 %, and a GC clamp (*see* **Note 15**). Avoid designing primers to sequences that are likely to form secondary structures (*see* **Note 16**) such as repetitive or palindromic sequences and those that are likely to amplify nonspecific products such as conserved sequence domains (*see* **Note 17**).

3.2 PCR

1. Mix PCR components (*see* **Notes 18–21**) as per Table 1 for a 1× PCR mastermix (*see* **Note 20**) with a final volume of 10 μl.
2. Aliquot 8 μl of mastermix into PCR vessel.
3. Add 2 μl DNA template (typically 20–150 ng) (*see* **Note 23**) to the PCR mix. Include controls in triplicate for the following: negative control (no DNA added) and positive controls for each expected genotype and a heterozygous sample (*see* **Note 24**).
4. Seal the plate with an optically clear seal (*see* **Note 25**).

Table 1
PCR components (*see* Note 22)

Reagent	Stock concentration	Final concentration	μl for 1×
DNA polymerase	5 U/μl	0.25 U	0.05
PCR buffer	10×	1×	1
dNTP	2 mM	0.2 mM	0.2
$MgCl_2$	50 mM	1–3 mM	0.3–0.6
Forward/reverse primer mix	10 μM	0.3 μM	0.3
SYTO®9	50 μM	1.5 μM	0.3
BSA	10 mg/ml	0.1 mg/ml	0.1
Water	–	–	To 8 μl

5. Cycle the PCR in a thermal cycler (*see* **Note 26**) with initial denaturation at 95 °C 10 min, 35 cycles of 94 °C 10–30 s, 60 °C 30 s, 72 °C 15 s, final extension at 72 °C 5 min (*see* **Note 27**).

3.3 HRM

1. Perform melting analysis on a Real-Time PCR system or an HRM instrument following the instrument protocol/recommendations. As a general guide the following HRM conditions can be used on the LightCycler® 480 Real Time PCR System (Roche): 95 °C for 10 s, 40 °C for 1 min (*see* **Note 28**), *Tm* –2 °C for 5 s (*see* **Note 29**), continuous acquisition mode up to 95 °C with 25 data acquisitions per degree change, and final cooling of 40 °C for 5 s.
2. Analyze melt data with instrument software (*see* **Note 30**). Analysis typically involves normalizing the data by setting pre- and post-melt regions to define the region of analysis (Fig. 1). The software defines the "pre-melt" region as 100 % fluorescence and assumes that all DNA samples are double stranded. The "post-melt" region is where all samples are single stranded and have no fluorescence. Ensure negative controls are negative and that replicated positive control samples are very similar in melt profiles (*see* **Note 31**). Interpret normalized melt plots and difference plots to distinguish genetic variants and determine sample genotypes. Normalized melt plots are interpreted as a change in melt curve shape, whereas difference plots depict differences in fluorescence relative to a selected sample.

4 Notes

1. Various companies sell ready-made HRM-PCR master mixes, e.g., LightCycler® 480 High Resolution Melting Master (Roche), MeltDoctor™ HRM Master Mix (Life Technologies) that streamline reaction setup and reduce reaction variability.
2. HRM compares the melt profiles of amplicons from independent PCRs. Therefore variation in sample preparation, sample concentration, and sample salt content can result in unwanted reaction-to-reaction variability. To minimize variability, use a standardized highly robust DNA extraction protocol for all samples, normalize the sample DNA concentration and use a set amount of DNA template in each PCR for the best results.
3. Hot-Start DNA polymerases are reported to improve reaction specificity by eliminating the presence of nonspecific products such as primer–dimers and mis-primed products.
4. dsDNA binding dyes are photosensitive and should be protected from light. SYTO®9 stain stock tubes should be covered in foil or stored in lightproof containers. When practically possible, PCR mastermix tubes and reaction vials (such as PCR plates) should be covered in foil or have limited light exposure.

Table 2
SNP classes

Class	Base change	Tm melt curve shift
1	C>T, T>C, G>A, A>G	Large
2	C>A, A>C, G>T, T>G	
3	C>G, G>C	
4	A>T, T>A	Very small

5. SYTO®9 stain adheres to glass, therefore, only use plastic tubes when diluting the stock.
6. SYTO®9 stain concentrated stock is stored in DMSO making it highly permeable to cell membranes. Appropriate personal protective equipment should be used when handling the stock solution to prevent the nucleic acid stain from entering cells.
7. No mutagenicity or toxicity data is available for SYTO®9 dye (www.invitrogen.com). However, since it binds to nucleic acids it should be treated as a potential mutagen. Therefore caution should be used when handling any solution containing SYTO®9 and appropriate personal protective equipment and waste disposal procedures should be followed.
8. Water should be prepared by purifying deionized water to 18 MΩ cm at 25 °C.
9. White PCR plates are advantageous over clear PCR plates since they reduce noise from fluorescent cross talk between wells and reflect light within a well for higher fluorescent signals.
10. Optically clear seals are effective in transmitting light and have minimal autofluorescence.
11. Free web-based primer design software include Primer3 http://frodo.wi.mit.edu/primer3/ [10], BatchPrimer3 http://batchprimer3.bioinformatics.ucdavis.edu/ [11], and NetPrimer http://www.premierbiosoft.com/netprimer/.
12. Typically multiple sets of primers need to be designed for an assay to determine which pair of primers is more efficient.
13. PCR primers can be synthesized to PCR/sequencing grade. If an assay is troublesome, sometimes HPLC purified primers can improve results.
14. The change in fluorescent intensity conferred by a sequence variant is easier to detect in short amplicons compared to long amplicons. This is particularly important when genotyping a fragment that only contains 1 SNP. SNP are categorized into classes depending on the base change conversion (Table 2). The class system reflects the number of hydrogen bonds

required to be broken and this is correlated with the expected observed difference in a melt curve [7].

15. A GC clamp has two or three out of the last five bases of the primer as guanines or cytosine.
16. Secondary structures such as hairpins, cross dimers or self-dimers, can greatly reduce the efficiency of PCR amplification and hence amplicon yield.
17. If possible use BLAST (http://blast.ncbi.nlm.nih.gov/) to determine if the primer sequences are specific. For polyploid species, where design of genome-specific primers can be difficult, conserved primers amplifying homoeologous sequences of the same length can be used to detect polymorphisms since the HRM analysis is sensitive to allele dosage.
18. Ensure that all reaction components are adequately mixed and collected before use.
19. Accurate pipetting must be used when setting up PCR to ensure that all reactions are of equal volume and contain the same concentration of reaction components. Differences in reaction components, in particular salt concentrations, can result in differences in DNA melting behaviour.
20. Incorporation of foreign bodies into PCR reactions such as dust or wool fragments should be avoided as these moieties can result in aberrant fluorescent data acquisition.
21. Scale up mastermix for more reactions.
22. $MgCl_2$ concentration greatly influences the melting behaviour of a DNA amplicon. Therefore the optimum $MgCl_2$ concentration for an assay should be empirically determined in a pre-assay of new primers.
23. The optimum amount of DNA template required for the PCR reaction varies depending on the genome size of the species under investigation.
24. An artificial heterozygous sample can be made by combining equal amounts of genomic DNA from the opposing genotypes under investigation. The heterozygous sample is an important control since heteroduplexes (formed from a heterozygous sample) have very different melting properties to homoduplexes. Differences in melt curve profiles between this sample and the other homozygous samples will indicate that a polymorphism is present that cannot be detected in the homozygous melt curves under the conditions used.
25. Ensure that PCR plates are adequately sealed with an optical clear seal to avoid evaporation of PCR. Avoid putting fingerprints or any other foreign object on the optically clear PCR seal as it can interfere with data acquisition during HRM.
26. PCR cycling can be performed on the Real-Time PCR machine that will acquire HRM data, or it can be done on a normal

PCR machine. It is advantageous to cycle on the Real-Time PCR machine to determine if amplification has been uniform across all samples. A measure of the uniformity is represented by a value often termed as the "cycle threshold" or the "crossing point". This value and the algorithm used to calculate it will vary depending on the instrument and the software used.

27. Optimum PCR reaction components and PCR cycling need to reproducibly generate robust products of high yield and specificity suitable for HRM analysis. Agarose gel electrophoresis is typically used to determine if a strong, single, and specific product of the correct size has been amplified, that contains little or no primer dimer. The concentrations of $MgCl_2$, primers and DNA can be altered to optimize an HRM assay. The thermal cycling conditions can also be altered to optimize an HRM assay.
28. Cooling to 40 °C aids heteroduplex formation and reassociation of all DNA products.
29. Choose the temperature to be 2 °C below primer-specific temperature.
30. Software packages vary for each melting instrument. Specific analysis details can be found in specific instrument manuals.
31. Discrepancies in positive control replicates indicate inaccuracies in pipetting and reaction component concentrations.

References

1. Wittwer CT, Reed GH, Gundry CN et al (2003) High-resolution genotyping by amplicon melting analysis using LCGreen. Clin Chem 49:853–860
2. Lochlainn SO, Amoah S, Graham NS et al (2011) High Resolution Melt (HRM) analysis is an efficient tool to genotype EMS mutants in complex crop genomes. Plant Methods 7:43
3. Mackay JF, Wright CD, Bonfiglioli RG (2008) A new approach to varietal identification in plants by microsatellite high resolution melting analysis: application to the verification of grapevine and olive cultivars. Plant Methods 4:8
4. Dong C, Vincent K, Sharp P (2009) Simultaneous mutation detection of three homoeologous genes in wheat by high resolution melting analysis and mutation surveyor®. BMC Plant Biol 9:143
5. Huynh B-L, Mather D, Schreiber A et al (2012) Clusters of genes encoding fructan biosynthesizing enzymes in wheat and barley. Plant Mol Biol 80:299–314
6. Reed GH, Kent JO, Wittwer CT (2007) High-resolution DNA melting analysis for simple and efficient molecular diagnostics. Pharmacogenomics 8:597–608
7. Liew M, Pryor R, Palais R et al (2004) Genotyping of single-nucleotide polymorphisms by high-resolution melting of small amplicons. Clin Chem 50:1156–1164
8. Paux E, Faure S, Choulet F et al (2010) Insertion site-based polymorphism markers open new perspectives for genome saturation and marker-assisted selection in wheat. Plant Biotechnol J 8:196–210
9. Monis PT, Giglio S, Saint CP (2005) Comparison of SYTO9 and SYBR Green I for real-time polymerase chain reaction and investigation of the effect of dye concentration on amplification and DNA melting curve analysis. Anal Biochem 340:24–34
10. Rozen S, Skaletsky HJ (2000) Primer3 on the WWW for general users and for biologist programmers. In: Krawetz S, Misener S (eds) Bioinformatics methods and protocols: methods in molecular biology. Humana, Totowa, NJ, pp 365–386
11. You F, Huo N, Gu Y, M-c L, Ma Y, Hane D, Lazo G, Dvorak J, Anderson O (2008) BatchPrimer3: a high throughput web application for PCR and sequencing primer design. BMC Bioinforma 9:253

Chapter 6

Bi-Allelic SNP Genotyping Using the TaqMan® Assay

John Woodward

Abstract

With TaqMan® technology allele-specific probes are utilized for quick and reliable genotyping of known polymorphic sites. TaqMan assays are robust in genotyping multiple variant types, including single nucleotide polymorphisms, insertions/deletions, and presence/absence variants. To query a single bi-allelic polymorphism, two TaqMan probes labeled with distinct fluorophores are designed such that they hybridize to different alleles during PCR-based amplification of a surrounding target region. During the primer extension phase of PCR, the 5′–3′ exonuclease activity of *Taq* polymerase cleaves and releases the fluorophores from bound probes. At the end of PCR, the emission intensity of each fluorophore is measured and allele determination at the queried site can be made.

Key words 5′ Nuclease assay, Applied biosystems, 6-FAM™, VIC®, Single nucleotide polymorphism, Genotyping

1 Introduction

Genetic variation in the form of single nucleotide polymorphisms (SNP), insertions/deletions, presence/absence variants, and copy-number variations greatly contribute to the phenotypic diversity observed within a species [1]. While historical breeding practices relied on phenotypic observations to infer genetic inheritance, current molecular breeding strategies utilize assays to genetic variation, or genetic markers, to predict phenotypes [2–4]. Such practices have revolutionized plant breeding and have accelerated genetic gain [2–4]. Marker-assisted selection (MAS) of genomic loci conditioning phenotypic effect, for example, has greatly expedited introgression of desirable traits into elite breeding material [3, 4].

Single-nucleotide polymorphisms are common variants found throughout the genome that are highly amendable to robust assay development [5]. SNP have been effectively used in a wide range of breeding applications including quantitative trait locus mapping, genetic map construction, genome-wide association analysis, MAS, and pedigree and population structure analysis [2–5].

Delphine Fleury and Ryan Whitford (eds.), *Crop Breeding: Methods and Protocols*, Methods in Molecular Biology, vol. 1145, DOI 10.1007/978-1-4939-0446-4_6, © Springer Science+Business Media New York 2014

The 5′ nuclease allelic discrimination assay, or TaqMan® assay, is a fast and cost-effective method for low-density SNP genotyping [6, 7]. TaqMan® is a registered trademark of Roche Molecular Systems, Inc. While TaqMan can be used to assay many types of genetic variation, this chapter focuses on bi-allelic SNP genotyping.

A single TaqMan assay utilizes two probes that specifically hybridize to each SNP allele during PCR amplification of a flanking target sequence. The probes are 13–18 bp oligonucleotides dual-labeled with a florescence quencher at the 3′ end and either the FAM™ (6-carboxyfluorescein) or VIC® (4,7,2′-trichloro-7′-phenyl-6-carboxyfluorescein) fluorophore at the 5′ end. A minor groove binder (e.g., dihydrocyclopyrroloindole tripeptide [DPI3]) is incorporated at the 3′ end of TaqMan probes to increase the stability and specificity of probe hybridization [8]. During the extension step of PCR, *Taq* polymerase adds nucleotides from primers flanking the polymorphism of interest. When the *Taq* polymerase reaches a bound TaqMan probe, the 5′–3′ exonuclease activity of *Taq* cleaves the FAM or VIC fluorophore relieving the quenched state. Single base-pair mismatches between the probe and nontarget alleles destabilize probe hybridization, limiting nonspecific probe binding and subsequent cleavage of fluorophores. Following PCR, the distinct emission wavelengths of FAM (518 nm) and VIC (554 nm) are captured, which allows for allele determination from the analyzed DNA sample. A sample that is homozygous for one allele will have fluorescence from only the respective FAM or VIC fluorophore, while a sample that is heterozygous at the analyzed locus will have both FAM and VIC fluorescence.

2 Materials

2.1 Tools, Instruments, and Software

1. Personal protective equipment (e.g., gloves, lab coat, safety glasses).
2. Plastic microcentrifuge tubes (1.5, 2.0 mL) for large master mix preparation.
3. Table-top centrifuge with rotor for plates.
4. Micropipettes (0.5–200 μl).
5. 8- or 12-channel micropipettes (0.5–200 μl).
6. Optical 96- or 384-well PCR plates and adhesive seals: MicroAmp® 96-Well Plates (Invitrogen); MicroAmp® 384-Well Plates (Invitrogen); MicroAmp® Optical Adhesive Films (Invitrogen).
7. 96- or 384-well PCR thermocycler. The following instruments are recommended by Applied Biosystems (AB): 9800 Fast Thermal Cycler, using the 9700/9600 emulation mode, or GeneAmp® PCR System 9700 (*see* **Notes 1** and **2**).

8. Real-time PCR System. The following instruments are recommended by Applied Biosystems: AB 7900 HT Fast Real-Time PCR System, AB 7500 Fast Real-Time PCR System, ABI PRISM® 7000 Sequence Detection System, AB 7300 or 7500 Real-Time PCR System (*see* **Notes 1** and **2**).
9. Software: Sequence Detection Software (Applied Biosystems), Primer Express® Software Version 3.0 (Applied Biosystems), Primer3 (http://frodo.wi.mit.edu/).

2.2 Reagents

1. Quantified template DNA (4 ng/μL), store at −20 °C.
2. 2× KlearKall Mastermix + ROX (*see* **Note 3**), store at −20 °C.
3. Nuclease-free water.
4. Forward and Reverse Primers (*see* **Note 3**), store at −20 °C.
5. Custom FAM™ and VIC® labeled TaqMan® probes (Applied Biosystems).

3 Methods

The TaqMan reaction requires two primers and two probes in a standard PCR reaction mix. Allele determination at a polymorphic site within a DNA sample can be made at the end of PCR using a fluorescence reader and genotyping software. The following describes custom designed assays; however, Applied Biosystems offers pre-designed SNP assays in certain organisms, and also has proprietary software for automated TaqMan assay design (www.appliedbiosystems.com).

3.1 TaqMan Assay Primer Design

To successfully query a polymorphism using the TaqMan assay, a target region that includes the SNP must be amplified using two PCR primers. Primers can be generated using standard publically available primer design software such as Primer3 (http://frodo.wi.mit.edu/; 9). Rules and guidelines for successful primer design are below [7, 9]:

1. The forward and reverse primers must flank the target polymorphism.
2. Primers should ideally amplify a 70–150 bp fragment.
3. Melting temperature (T_m) of the primer pair: 59–62 °C (optimum: 60 °C).
4. Primer size: 18–28 bp (optimum: 20 bp).
5. Total GC content: 30–70 %.
6. The max T_m difference between the forward and reverse primer should be 1 °C.
7. Each primer should have less than five repeating nucleotides in a row.

8. The total number of G and C in the last five nucleotides at the 3′ end of the primer ideally should not exceed two.
9. At least one and ideally both primers should hybridize to only one region in the genome so that a single unique amplicon is generated during PCR (*see* **Note 4**).
10. Primer3 default parameters can be used for other design criteria.
11. Primers can be ordered using oligo suppliers such as IDT (www.idtdna.com).

3.2 TaqMan Assay Probe Design

Two probes that minimally vary at the polymorphic site and are labeled with different fluorophores are required in the TaqMan assay. Although the probes can be offset around the polymorphic site, they should be designed so that they bind the complementary target genomic region with approximately equal efficiencies. This can be accomplished by designing under a strict set of parameters that restrict the hybridization properties of the probes. Probes can be designed manually using software such as Primer Express® Software Version 3.0 from Applied Biosystems or automatically using custom design tools from ABI (https://www5.invitrogen.com/custom-genomic-products/tools/gene-expression/). Rules and guidelines for successful probe design are below [7]:

1. Assign the 6-FAM™ fluorophore to one of the probes and VIC® to other probe within the assay (*see* **Note 5**).
2. Probe length: 13–18 bp.
3. T_m: 65–67 °C.
4. GC content: 30–80 %.
5. Attempt to place the SNP in the middle 1/3 of the design sequence (away from the 5′ end).
6. Avoid homopolymers of four or more; avoid three consecutive G.
7. Attempt to have C:G ratio > 1.
8. Avoid excessive G or C at the 3′ end (three or more); this will help prevent nonspecific binding.
9. Avoid G in the first or second bp.
10. Attempt to have only one SNP per probe, unless adjacent SNP are in linkage disequilibrium.
11. TaqMan probes can be ordered from Applied Biosystems.

3.3 Setting Up the PCR Assay

During PCR amplification of a target region, fluorophores are cleaved and released from probes bound at the target site. There are many choices of *Taq* and PCR master mixes available for PCR. The following protocol is performed with KlearKall Matermix + ROX (http://www.lgcgenomics.com/klearkall-mastermix) (*see* **Note 6**). During PCR setup and subsequent analysis, attempt to limit exposure to light as it may dampen the fluorescence.

Table 1
PCR mix for TaqMan SNP assay

Component (starting concentration)	Volume (μL)/well 96-well plate	Volume (μL)/well 384-well plate	NTC—volume (μL)/well 384-well plate
DNA (16 ng dried)	–	–	No DNA
Nuclease-free water	9.66	2.415	2.415
KlearKall Mastermix—2×	10	2.5	2.5
Forward primer (100 μM)	0.15	0.0375	0.0375
Reverse primer (100 μM)	0.15	0.0375	0.0375
Probe 1 (100 μM)	0.02	0.005	0.005
Probe 2 (100 μM)	0.02	0.005	0.005
Total	20	5	5

1. Set up the PCR reaction in a PCR 96- or 384-well optical PCR plate.
2. Add the following PCR mix to 16 ng of dried DNA as described in Table 1. Also, prepare at least two no template controls (NTC) in each plate by pipeting the reaction mix into empty wells. Control DNA samples, which are known to be polymorphic at a targeted SNP, can be included as quality controls. Make sure to thoroughly mix components prior to use (*see* **Note 7**).
3. Mix reactions by vortexing and collect by brief centrifugation.
4. Place reactions in a standard thermocycler, heat-start the reaction at 95 °C for 10 min, and run 40 cycles of the following conditions: 95 °C for 30 s (denature), 60 °C for 60 s (anneal and extend).
5. After PCR thermocycling cool to 4 °C and briefly centrifuge reactions to collect liquid.
6. Keep reactions at 4 °C in the dark or store at −20 °C until fluorescence measurement.

3.4 Allele Determination

At the end of PCR, allele determination can be made by measuring the ratio of the fluorescence signals for each of the reporter dyes. An Applied Biosystems Real-Time PCR System can be used to measure fluorescence and the Applied Biosystems Sequence Detection System (SDS) software can be used to plot and analyze the signal output (*see* **Note 1**). For SDS software on the 7900 HT Fast Real-Time PCR System the following applies:

1. Open the SDS Software, create a new document, and apply "Allelic Discrimination" in the assay drop-down and the appropriate PCR plate.

2. Set-up detector information using the Detector Manager dialog box. Add FAM, VIC detectors with specific colors and Create Markers.
3. Assign marker information and apply markers to the plate document:
   (a) Use "Marker Manager" under "Tools" to add markers and apply markers to the allelic discrimination plate.
   (b) Select "File" and "New" to access the Add Detector dialog box.
   (c) Select, Name, and Create the FAM and VIC detectors. Select none for quencher. Click "OK" to save detectors.
   (d) Under the "Marker Manager" Dialog select "Create Marker," type a name and add the detectors you created above.
   (e) Click "Copy To Plate Document" and then click "Done" within the Marker Manager Dialog.
   (f) Use the Allelic Discrimination dialog box to identify wells to be analyzed; use Ctrl and Shift keys to select wells.
   (g) Add FAM and VIC markers to the selected wells by checking boxes within the marker inspector.
   (h) Apply tasks by selecting wells and assigning unknown or NTC within the well inspector.
   (i) Assign ROX as the Passive reference.
4. Perform a plate read by selecting the "Instrument" tab at the Allelic Discrimination dialog box:
   (a) The plate read is performed at 60 °C.
   (b) Set the volume to reaction plate volume (5 or 20 μL).
   (c) Save File and select "Post Read".
5. Perform plot analysis and make allele determinations by selecting clusters.

4 Notes

1. Many different thermocycling instruments can be reliably used to perform the PCR step of the TaqMan assay. It is always useful to include positive controls within an experiment and perform technical and biological replications to ensure results and troubleshoot. Furthermore, fluorescence measurements can be made using many standard plate readers, such as the Tecan Safire (www.Tecan.com) or the PHERAstar (http://www.bmglabtech.com). Other commercial software packages, such as BioNumerics (Applied Maths) or Kraken^{TM} (http://www.lgcgenomics.com), can be used for analysis of the fluorescence

output and genotyping. Please review the instrument and software-specific procedures to perform genotypic analyses using other systems.

2. Applied Biosystems has many help documents online for performing SNP assays (www.appliedbiosystems.com): TaqMan SNP Genotyping Assays Protocol (http://www3.appliedbiosystems.com/cms/groups/mcb_support/documents/generaldocuments/cms_042998.pdf); 7900 HT Allelic Discrimination Getting Started Guide (PN 4364015); Applied Biosystems 7300/7500/7500 Fast Real-Time PCR System Allelic Discrimination Getting Started Guide (PN 4347822); 7000 System—ABI PRISM® 7000 Sequence Detection System User Guide (PN 4330228).

3. Multiple PCR master mixes containing *Taq* polymerase can be used to perform the TaqMan assay. Applied Biosystems suggests using the TaqMan® Genotyping Master Mix containing AmpliTaq Gold® and ordering their TaqMan® SNP genotyping assay, which contains the target forward and reverse primers and the FAM and VIC Probes in a single assay tube. However, manually designing and ordering separate primers and probes as described offers additional flexibility to optimize each TaqMan SNP assay.

4. Often it is useful to align primer sequences against a genomic reference sequence to verify that the primer pair will only amplify a single target amplicon. This may be difficult in species without reference sequence or in highly conserved polyploid genomes.

5. FAM and VIC are good opposing fluorophores as they do not have overlapping excitation/emission maxima.

6. ROX™, which is a glycine conjugate of 5-carboxy-X-rhodamine, succinimidyl ester, is a passive reference dye used to normalize the fluorescence of the reporters across samples (http://tools.invitrogen.com/content/sfs/manuals/rox_referencedye_man.pdf). ROX normalization results in increased data precision by partially controlling experimental variation. The excitation and emission wavelength maxima of the reference dye are ~575 nm and ~605 nm, respectively.

7. A large PCR master mix for multiple reactions can be created by multiplying each volume for a reaction component by $n+4$ samples (where $n=$ total number of DNA samples + NTC and Positive controls) and combining all components into a single 1.5 or 2.0 mL tube. Creation of a larger master mix volume than expected for n samples is often required to account for variation in pipeting accuracy. An aliquot of 20 or 5 μL can then be added to dried-down DNA in the 96 or 384 plates, respectively.

References

1. Kaeppler S (2012) Heterosis: many genes, many mechanisms—end the search for an undiscovered unifying theory. ISRN Botany, 2012. doi:10.5402/2012/682824
2. Tester M, Langridge P (2010) Breeding technologies to increase crop production in a changing world. Science 327:818–822
3. Eathington SR, Crosbie TM, Edwards MD et al (2007) Molecular markers in a commercial breeding program. Crop Sci 47(3):154
4. Moose SP, Mumm RH (2008) Molecular plant breeding as the foundation for 21st century crop improvement. Plant Physiol 147:969–977
5. Rafalski A (2002) Applications of single nucleotide polymorphisms in crop genetics. Curr Opin Plant Biol 5:94–100
6. Holland PM, Abramson RD, Watson R et al (1991) Detection of specific polymerase chain reaction product by utilizing the 5′–3′ exonuclease activity of *Thermus aquaticus* DNA polymerase. PNAS 88:7276–7280
7. Livak KJ (1999) Allelic discrimination using fluorogenic probes and the 5′ nuclease assay. Genet Anal 14:143–149
8. Kutyavin IV, Afonina IA, Mills A et al (2000) 3′-minor groove binder-DNA probes increase sequence specificity at PCR extension temperatures. Nucleic Acids Res 28:655–661
9. Rozen S, Skaletsky H (2000) Primer3 on the WWW for general users and for biologist programmers. Methods Mol Biol 132:365–386

Chapter 7

SNP Genotyping: The KASP Assay

Chunlin He, John Holme, and Jeffrey Anthony

Abstract

The KASP genotyping assay utilizes a unique form of competitive allele-specific PCR combined with a novel, homogeneous, fluorescence-based reporting system for the identification and measurement of genetic variation occurring at the nucleotide level to detect single nucleotide polymorphisms (SNPs) or inserts and deletions (InDels). The KASP technology is suitable for use on a variety of equipment platforms and provides flexibility in terms of the number of SNPs and the number of samples able to be analyzed. The KASP chemistry functions equally well in 96-, 384-, and 1,536-well microtiter plate formats and has been utilized over many years in large and small laboratories by users across the fields of human, animal, and plant genetics.

Key words SNPs, Genotyping platform, KASP High throughput, Genetic variation

1 Introduction

Single nucleotide polymorphism (SNP) markers, the most common form of genetic variation among individuals, are the most recently developed DNA marker technology and are considered the preferred marker system for the study of complex genetic traits, such as genome-wide association studies, and for breeding applications. Their abundance in the genomes of all organisms, amenability to low-cost, high-throughput genotyping and the relatively low mutation rate compared to other markers make SNP markers ideal candidates for studies aimed at the creation of improved agricultural organisms and providing solutions to scientific problems [1–8]. The measurement of genetic variation caused by SNPs starts with the identification or determination of the genotypes of the particular individuals of the same species, namely "genotyping".

The KASP assay utilizes a novel homogeneous fluorescent genotyping system, provides good flexibility and its chemistry functions equally well in 96-, 384-, and 1,536-well plate formats [9, 10]. Combined with assay miniaturization in 1,536-well PCR formats KASP is able to deliver high levels of flexibility in generating data

Delphine Fleury and Ryan Whitford (eds.), *Crop Breeding: Methods and Protocols*, Methods in Molecular Biology, vol. 1145, DOI 10.1007/978-1-4939-0446-4_7, © Springer Science+Business Media New York 2014

sets from 1 SNP over as few as 22 samples, i.e., only 22 data points, to thousands of SNPs over thousands of samples generating millions of data points in a single day [9]. KASP has been utilized over many years by large and small laboratories to drive research targeting the genetic improvement of animals [9] and field crops [4, 5]. KASP was chosen by the Generation Challenge Programme (www.generationcp.org) as the method of choice for the SNP genotyping services offered by its Integrated Breeding Platform (www.integratedbreeding.net).

The constituent oligonucleotides necessary for the mechanism of action behind KASP are the following :

- Two allele-specific primers (one for each SNP allele). Each primer contains a unique unlabelled tail sequence at the 5′ end.
- One common (reverse) primer.
- Two 5′ fluor-labeled oligos, one labeled with FAM, one with HEX. These oligo sequences are designed to interact with the sequences of the tails of the allele-specific primers.
- Two oligos, with quenchers bound at the 3′ ends. These oligo sequences are complementary to those of the fluor-labeled oligos (and therefore also complementary to the tails of the allele-specific primers). These quenched oligos therefore bind their fluor-labeled complements and all fluorescent signals are quenched until required.

In the initial stage of PCR, the appropriate allele-specific primer binds to its complementary region directly upstream of the SNP (with the 3′ end of the primer positioned at the SNP nucleotide) (Fig. 1). The common reverse primer also binds and PCR proceeds, with the allele-specific primer becoming incorporated into the template. During this phase, the fluor-labeled oligos remain bound to their quencher-labeled complementary oligos, and no fluorescent signal is generated. As PCR proceeds further, one of the fluor-labeled oligos, corresponding to the amplified allele, is also incorporated into the template, and is hence no longer bound to its quencher-labeled complement. As the fluor is no longer quenched, the appropriate fluorescent signal is generated and detected by the usual means. If the genotype at a given SNP is homozygous, only one or the other of the possible fluorescent signals will be generated. If the individual is heterozygous, the result will be a mixed fluorescent signal.

2 Materials

2.1 The KASP Assay Constituents

The KASP reaction consists of the KASP assay mix (assay specific) and KASP Master mix (universal; used with any assay mix) which are combined with the DNA sample to be analyzed. All genotyp-

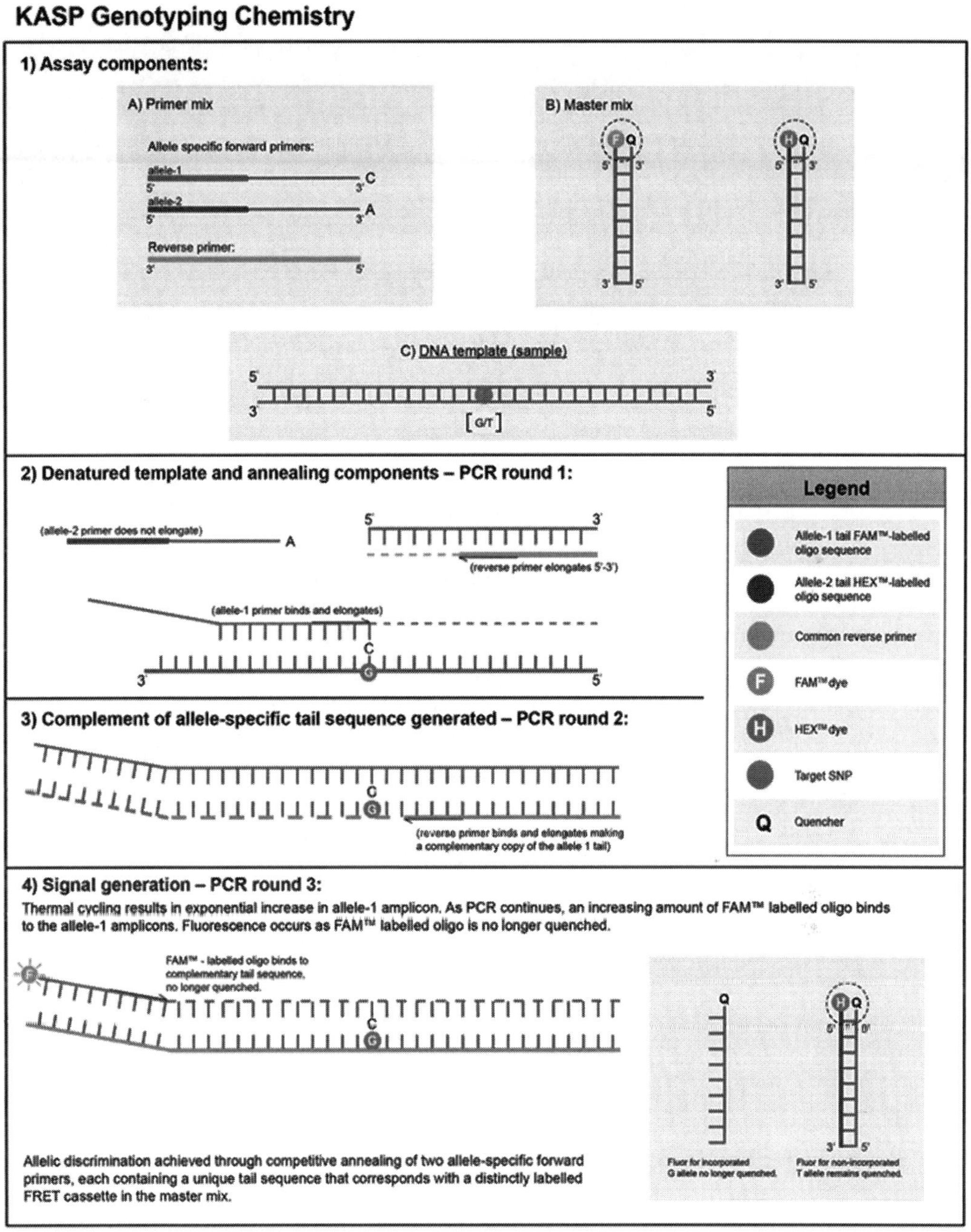

Fig. 1 An overview of KASP mechanism of action

ing solutions should be made with ultrapure water and analytical grade reagents then stored at room temperature or as indicated in the procedure. For optimized results it is important that users follow the protocol supplied with KASP reagents, including assay preparation and optimization as well as data capturing and analysis.

Table 1
Assembly of the constituent reagent volumes for KASP genotyping mix

Component	Wet-DNA method (μL)		Dry DNA method (μL)	
DNAs[a]	*2.5*	*5*	N/A	N/A
KASP Master mix (2×)	2.5	5	2.5	5
KASP assay mix (72×)	0.07	0.14	0.07	0.14
H_2O	N/A	N/A	2.5	5
Total reaction volume	5	10	5	10

[a]DNA samples are diluted to working concentration (*see* **Note 2**)

1. DNA samples can be prepared from a variety of DNA sources and technologies, including crude lysates. Generally, they all work well with the KASP assays if they meet the minimum requirements of quality (*see* **Note 1**) and quantity (*see* **Note 2**). Typically around 5 ng of DNAs is required for each PCR. DNA may be purified using any suitable technique. When a commercial purification technology is used, it is important to follow all the manufacturer's instructions. DNA can be diluted (*see* **Notes 3** and **4**) or used in dried form (*see* **Note 5**).
2. KASP assay mix can be ordered ready to use from the supplier (LGC Genomics). Primers are designed based on sequence data submitted by the user, containing the SNP/InDel to be targeted. The SNP/InDel-specific assays are available in two formats, either as in silico validated or fully validated formats (*see* **Note 6**). Primer design is achieved by importing SNP sequences into the Kraken™ software which generates the relevant oligo sequences of two competitive forward primers and a common reverse primer. Prepare the assay mix as per Table 1 and combine with KASP Master mix. Assay mix can be stored at 4 °C for 1–2 weeks, at −12 °C for about 1 year or indefinitely at −80 °C.
3. KASP Master mix is provided in a ready-to-use 2× format containing universal fluorescent reporting dyes FAM™ and HEX™ as well as Rhodamine X (ROX) dye as the passive reference. Provided separately to the KASP Master mix are a stock solution of 50 mM $MgCl_2$ and dimethyl sulfoxide (DMSO) (*see* **Note 7**).

2.2 Equipment

The KASP system does not require customers to purchase all necessary equipment; only major equipment can be purchased from LGC if required or existing lab equipment can be adapted to fit the KASP system. The assay process works with standard liquid handling, thermal cycling, and plate reading equipment that many labs may already have.

1. PCR microtiter plates. KASP genotyping can be performed in any plate well density but 96-, 384-, or 1,536-well microtiter plates are typically used for PCR.
2. Reagent dispensing equipment. A Meridian dispenser can be used for robotically dispensing reagent for the preparation of KASP assays. Alternatively, the process can be carried out manually with a suitable pipette, depending on plate type and the number of samples (*see* **Note 8**).
3. A plate sealer, such as Kube plate sealing instrument, Fusion3™ Laser Sealer or other adapted sealers, should be used for sealing the PCR plates before they are put in waterbath for thermal cycling (*see* **Note 9**).
4. Thermal cycling instrumentation. The KASP chemistry can be used with any standard thermal cycler. Similar results can be obtained on Peltier block-based and waterbath-based thermal cyclers. When used in a high-throughput environment, the use of Hydrocycler™, a waterbath-based thermal cycling system, is recommended (*see* **Note 10**).
5. FRET capable plate reader. Most FRET-capable plate readers (with the relevant filter sets) can be used in conjunction with KASP. KASP uses the fluorophores FAM and HEX to distinguish genotypes. The passive reference dye ROX is also used to allow normalization of variations in signal caused by differences in well-to-well liquid volume (*see* **Note 11**).

3 Methods

All genotyping procedures should be carried out at room temperature unless otherwise indicated. Users should refer to the KASP SNP genotyping manual (http://www.lgcgenomics.com/genotyping/kasp-genotyping-reagents/). The KASP genotyping assay methodology is outlined in Fig. 2.

3.1 Prepare the DNA Plates

Array DNA samples into PCR microtiter plate. Create replicates of microtiter plates (MTP) with DNA samples by adding approximately 10 μg of DNAs to 96-well plates and 5 μg to 384- or 1,536-well plates, respectively. Based on the number of SNPs and samples required in the genotyping project, multiple copies of each MTP containing DNA to be analyzed can be created using a robotic platform such as LGC's RepliKator™ (*see* **Note 12**).

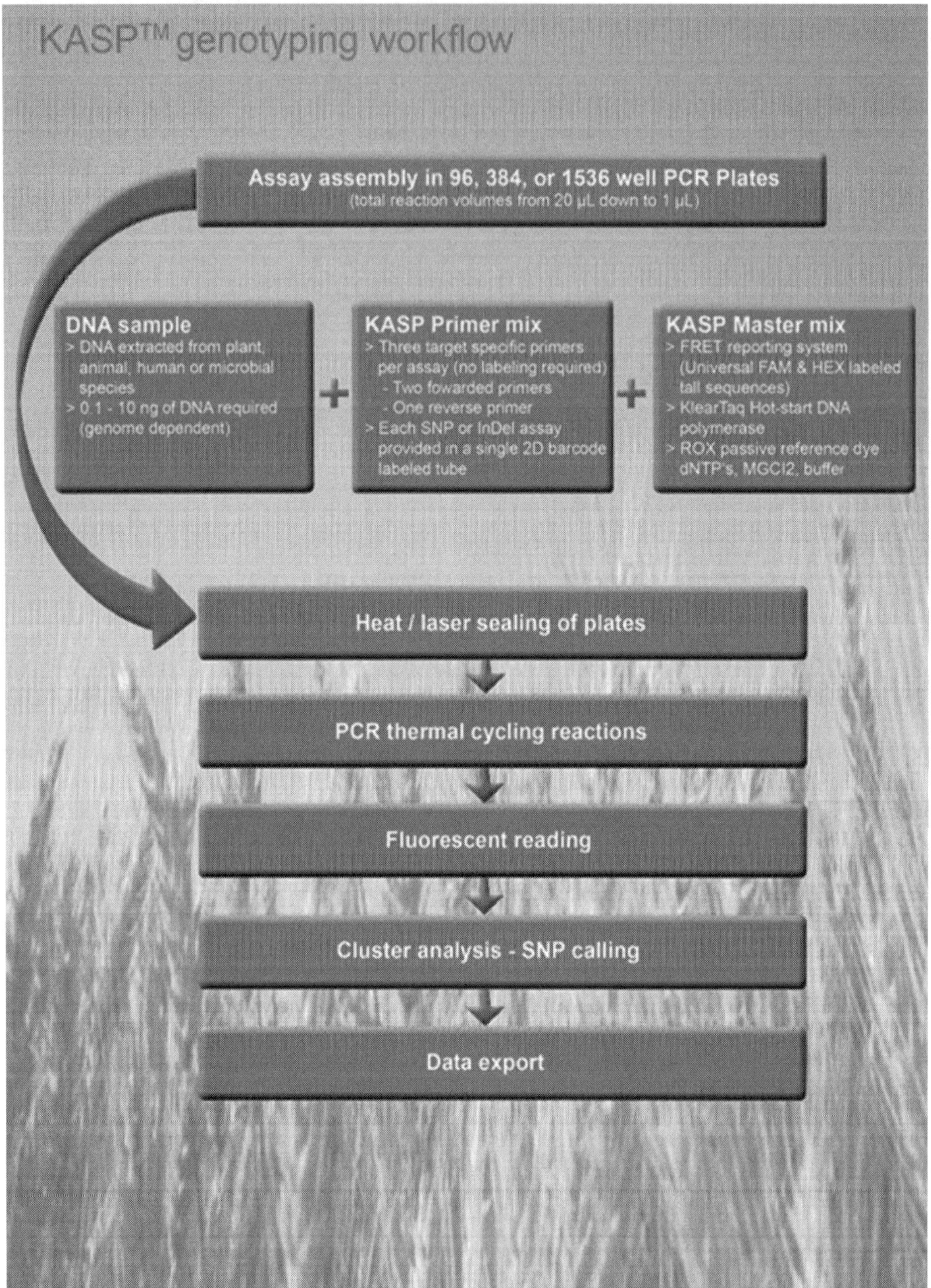

Fig. 2 The KASP workflow

Table 2
An example of KASP assay reagent bulk assembly for 60 PCR[a]

Component	Wet DNA method (μL) for 60 reactions		Dry DNA method (μL) for 60 reactions	
KASP Master mix (2×)	150	300	150	300
Assay mix (72×)	4.2	8.4	4.2	8.4
H_2O	N/A	N/A	150	300
Total reaction volume	5	10	5	10
Plate format	384-well plate	96-well plate	384-well plate	96-well plate
Total	300[b]	600[b]	300	600

[a]In this example, sufficient mix is provided for 56 DNA samples plus 4 no-template controls (NTCs), with an additional spare volume. The KASP assay reagents must be mixed prior to dispensing into the DNA samples

[b]Including the volume of DNA solution

Table 3
Thermal cycling conditions for the KASP genotyping system

Step	Description	Temperature (°C)	Time	# Cycles per step
1	Enzyme activation	94	15 min	1
2	Denature Annealing/elongation	94 61–55	20 s 60 s (drop 0.6 °C per cycle)	10
3	Denature Annealing/elongation	94 55	20 s 60 s	26

3.2 Prepare the KASP Assay

Table 2 describes relative volumes for combining the KASP Assay reagents. In reality, the pipetting volumes are not practical and hence a bulk mix should be assembled as described in Table 2.

1. Dispense KASP Assay reagents into DNA samples. Dispensing can be carried out robotically or manually with a suitable pipette, depending on the type of plates and the number of samples (*see* **Note 8**).
2. Seal the microtiter plate. The microtiter plate should be effectively sealed to prevent leakage and/or evaporation during thermal cycling. Creation of a perfect seal is an important element of the workflow. Use the Kube™ heat sealer to seal 96- or 384-well plates and the Fusion 3™ laser sealer welding system to seal the 384- or 1,536-well plates (*see* **Note 9**).
3. Run PCR using a thermal cycling instrument. The PCR thermocycling regime in Table 3 should be used for optimal generation of PCR products.

Table 4
Additional thermal cycling conditions for the KASP genotyping system

Step	Description	Temperature (°C)	Time (s)	# Cycles
1	Denature	94	20	3
	Annealing/elongation	57	60	

4. Read plate in fluorescent plate reader. Post PCR, the plates should be read in an appropriate plate reader. KASP utilizes the fluorophores FAM (excitation at 485 nm and emission at 520 nm) and HEX (excitation at 535 nm and emission at 556 nm) to differentiate the genotypes, and the passive reference dye ROX (excitation at 575 nm and emission at 610 nm) to normalize the variation. The FAM and HEX values should be divided by the ROX values.

3.3 Analyze Data

The data output from the fluorescent plate reader should be analyzed with a suitable software package (*see* **Note 13**). If clear genotyping data is not obtained, the plate should be thermally cycled further using the conditions above (Table 4) and read again. Further cycling and reading can be performed as required until tight genotyping clusters have been attained (*see* **Note 14**).

4 Notes

1. When DNA samples are extracted from plant tissues, polysaccharides and polyphenols potentially remain in the DNA samples which may interfere with PCR. It is necessary to reduce these contaminants, via titration with polyvinylpyrrolidone (PVP) to around 3 % (final concentration) in the DNA solution to bind polyphenols, which may improve the PCR. Alternatively, if PCR inhibitors are present, but the DNA concentration is high, sample dilution might be feasible to effectively dilute out the inhibitors with minimum adverse effect. DNA purification can be further optimized through the use of a commercial available preparation technology such as those described below, which include the Kleargene and sbeadex® technologies of LGC Genomics.
2. In order to obtain sufficient genotyping results to show clustering, it is suggested to have a minimum of 22 DNA samples. For the purpose of validating the genotyping results, at least two no-template controls (NTCs) should be included in each plate. When validating an assay with low allele frequency, inclusion of positive controls (DNA samples with known genotypes) is recommended. It is recommended to use approximately 10 ng DNA for 96-well plates and 5 ng DNA for 384- and

1,536-well plates. When determining DNA concentration, it is to be expected that a lower figure will generally be obtained with Picogreen than with spectrophotometry.

3. Actual working concentrations of DNA used in KASP are dependent on genome size and complexity and the use of a greater mass of DNA per reaction is generally recommended if the organism has a larger genome. In addition, both high and low DNA concentrations can cause problems with poor or no PCR amplification. When DNA concentration is lower than expected, it can be addressed to some extent by extra thermal cycling for additional amplification. When DNA concentration is too high, it is suggested to dilute the DNA samples such that any PCR-inhibiting contaminants will also be diluted to non-problematic levels (though DNA concentration should obviously remain sufficiently high to allow PCR).

4. If DNA samples are dissolved in TE buffer (which contains EDTA), it is recommended to add extra Mg^{2+} as EDTA will chelate Mg^{2+} causing insufficient Mg^{2+} concentration. The amount of Mg^{2+} added should stoichiometrically reflect the EDTA added, e.g., if the TE buffer used for DNA dissolution contained 1 mM EDTA, an extra 1 mM Mg^{2+} should be added to the KASP mix to compensate. However, the current iteration of KASP has been developed to be less sensitive to Mg^{2+} concentration and hence to EDTA concentration. In practice a small subset of the DNA samples should be tested with a brief Mg^{2+} titration prior to commencement of a large genotyping project.

5. For a large-scale genotyping, it is recommended to prepare PCR plates containing DNA samples in advance and dry them by quickly centrifuging the plates and then placing the plates in a drying oven at 60 °C for approximately 1 h, or until dry. The use of dried DNA samples in plates can allow preparation of many sample plates in advance without the concern of sample evaporation which would affect the final reagent concentration of the PCR.

6. KASP assay mix is provided by LGC in two formats: KASP by Design (KBD) and KASP on Demand (KOD). KBD designs and provides un-validated assays but at a reduced price, while the KOD option supplies optimized assays guaranteed to produce good genotyping data in any laboratory setting. Sequence data for the SNP/InDel to be targeted is submitted to LGC's assay design team who design and prepare the assay mix using the Kraken assay design software. Assay mix is then shipped to the customer laboratory in 2D barcoded tubes for incorporation into the assay in combination with the KASP Master mix and DNA sample. Upon receipt of the assay Mix tubes the reagents should be prepared as described in Table 5. When preparing KASP assay mix, ensure an equal amount of the

Table 5
Assembly of KASP primer mix

Component (stock concentration)	Volume (μL)	Final concentration (μM)	Notes
Allele-specific primer 1 (100 μM)	12	12	Forward primer, reports with FAM
Allele-specific primer 2 (100 μM)	12	12	Forward primer, reports with HEX
Common primer (100 μM)	30	30	Reverse primer
Tris–HCl (pH 8.3) (10 mM)	46		Analytical grade water can be used as an alternative
Total	100		

allele-specific primers is used, unequal amount of allele-specific primers can cause the heterozygous group to move towards one or the other of the two homozygous groups, making it difficult to call the individual genotypes. However, where the heterozygous group is shifted off-center, a titrative increase of the opposite allele-specific primer may alleviate the problem.

7. For large majority of KASP assays, the final optimal $MgCl_2$ concentration is 2.5 mM. However, for some assays in A/T-rich regions (>70 %), poor amplification can occur. When this issue is encountered, it is recommended to increase $MgCl_2$ concentration by 0.3 mM for a final concentration of 2.8 mM. On the contrary, assays in high G/C percentage regions (>70 %) may occasionally cause poor amplification; in this case it is suggested to run the assay with low $MgCl_2$ concentration (1.8 mM) or add additional 5–10 % DMSO to the final volume of the assay to improve the PCR. It is worth noting however that in general, even when genotyping areas of the genome with high or low G/C percentages, KASP will, generally, still work under standard conditions. For initial use of the KASP genotyping assay, users are recommended to request a free trial kit from LGC so that it can be determined if the plate reader and the thermal cycler are functioning properly with the chemistry. The KASP chemistry should not be used at final concentration other than 1× as the correct concentration of PCR reagents is essential for successful PCR amplification.
8. For high throughput assay dispensing, the Meridian Dispenser can be configured with a single or 8-channel dispensing head. The dispenser is able to dispense less than 0.5 μL to several milliliters per channel into any of the three types of MTPs. The unit includes a wash station to wash the tips to prevent clogging and cross-contamination.

9. The Kube plate sealing instrument is recommended for sealing PCR plates prior to PCR to guarantee a perfect hermetic seal. Optically clear seals should be used to minimize blocking of light signal transmission. For higher density MTPs, such as 1,536- or 3,456-well plate formats, the Fusion3™ Laser Sealer is recommended for the sealing without the use of heat.
10. The Hydrocycler™ is a high-throughput thermal cycler using water as a source for conducting energy to the PCR plates for rapid and uniform thermal transfer across all wells and plates. The transfer time for the Hydrocycler to automatically move the PCR plates from one temperature waterbath to the next is about 3 s. The Hydrocycler typically requires 1–1.5 h to complete all KASP PCR cycles.
11. KASP is an end-point genotyping system and fluorescent reads should be taken post-PCR. Such reading should preferably be carried out at ambient temperature but in any case below 40 °C. Use the PHERAstar plate reader (BMG) to read all three types of PCR plates, or use BMG Omega F plate reader for 96- or 384-well plates only.
12. DNA can be reformatted using the LGC's repliKator™ instrument to create multiple daughter plates from DNA source plates. Parent plates can be reformatted or "stamped" to create many copies of different microtiter plates including 96-, 384-, and 1,536-well plates as well as reformatting the 96-/384-well plates into 1,536-well plates. For a less automated solution the Kpette™ device includes a 384-well channel pipetting head to be used as either a 96- or 384-well pipettor with a pipetting volume ranging from 1 μL to 125 μL. The use of plate adaptors also enables reformatting of 96-well replicate plates to 96- or 384-well plates. By using tip guides, 384-well source plates can be rearrayed into 1,536-well plates.
13. LGC provides three levels of software capable of interpreting genotyping data. Other commercially available data interpretation packages may also be suitable for data analysis.
    - Kraken™: use any of the five modules as needed for the genotyping workflow including (1) KASP Assay Picker module; (2) Project Management module; (3) Sample Management and Tracking module; (4) Meridian Engine control interface; (5) Genotyping Data Analysis & Reporting Module.
    - KlusterCaller™: Part of Kraken software for data analysis with reduced functionality.
    - SNPviewer™: Used for graphical viewing of genotyping data.
14. For graphic viewing of genotypic data, if the plot shows too many genotyping groups, it is likely that a non-target polymorphism

may exist near the SNP of interest, thus it is not possible to determine the heterozygotes and homozygotes for the different clusters. Under this scenario, at least one of the primers has to be relocated and designed so that no unwanted polymorphism will arise. Alternatively a wobble base can be included in the primer at the site of the neighboring SNP. On the other hand, the cluster plot may show fewer groups than expected, if so, one of the allele-specific primers has to be checked and redesigned to make it function correctly. If too many samples do not demonstrate good amplification and thus are clustered around the origin, it is likely that the DNA quality of those samples is poor or the DNA has been poorly arrayed resulting in little/no DNA, and/or the reaction mix is poorly dispensed. In addition, incomplete sealing of the PCR plates can also result in poor amplification. For example, Generation Challenge Program (GCP) has funded and collaborated in more than 100 SNP genotyping projects with LGC Genomics (originally with KBioscience) in the past 3 years. Among the completed projects, the vast majority demonstrated good results with the success rate of assays ranging from 80 to 96 %. However, a couple of projects only had success rate of <70 %, which resulted from poor DNA quality and quantity.

Acknowledgments

We would like to thank Dr. Xavier Delannay from the Generation Challenge Program (GCP), c/o CIMMYT for reviewing the chapter and providing the valuable suggestions.

References

1. Perkel J (2008) SNP genotyping: six technologies that keyed a revolution. Nat Methods 5: 447–453
2. Ragoussis J (2006) Genotyping technologies for all. Drug Discov Today Tech 3:115–122
3. Jenkins S, Gibson N (2002) High-throughput SNP genotyping. Comp Funct Genom 3:57–66
4. Ribaut J-M, de Vicente MC, Delannay X (2010) Molecular breeding in developing countries: challenges and perspectives. Curr Opin Plant Biol 13:1–6
5. Delannay X, McLaren G, Ribaut JM (2012) Fostering molecular breeding in developing countries. Mol Breed 29:857–873
6. Nijman IJ, Kuipers S, Verheul M, Guryev V, Cuppen E (2008) A genome-wide SNP panel for mapping and association studies in the rat. BMC Genomics 9:95
7. Zhao Z, Boerwinkle E (2002) Neighboring-nucleotide effects on single nucleotide polymorphisms: a study of 2.6 million polymorphisms across the human genome. Genome Res 12:1679–1686
8. Zhang Z, Gerstein M (2003) Patterns of nucleotide substitution, insertion and deletion in the human genome inferred from pseudogenes. Nucleic Acids Res 31:5338–5348
9. Robinson P, Ganske F (2012) High speed FRET based SNP Genotyping Measurement on the PHERAstar, http://www.bmglabtech.com/application-notes/fret/snp-genotyping-fret-Kbioscience-160.cfm. Accessed 4 March 2014
10. Livak KJ, Flood SJ, Marmaro J, Giusti W, Deetz K (1995) Oligonucleotide with fluorescent dyes at opposite ends provide a quenched probe system useful for detecting PCR product and nucleic acid hybridization. Genome Res 4:357–362

Chapter 8

Rindsel: An R Package for Phenotypic and Molecular Selection Indices Used in Plant Breeding

Sergío Perez-Elizalde, Jesús J. Cerón-Rojas, José Crossa, Delphine Fleury, and Gregorio Alvarado

Abstract

Selection indices are estimates of the net genetic merit of the individual candidates for selection and are calculated based on phenotyping and molecular marker information collected on plants under selection in a breeding program. They reflect the breeding value of the plants and help breeders to choose the best ones for next generation. *Rindsel* is an R package that calculates phenotypic and molecular selection indices.

Key words Smith selection index, Eigen selection index method, Lande and Thompson index, Molecular Eigen selection index method

1 Introduction

Marker-assisted selection uses phenotypic and molecular markers information to select best performer lines at each step of a breeding program. Markers target either major loci controlling desirable traits such as flowering date or quantitative trait loci (QTL) for complex trait such as yield. While it is relatively easy to select the lines carrying the best alleles for 1–3 traits, identifying the best combination of alleles becomes more complicated when multiple loci are tracked simultaneously. Moreover phenotype data of lines under selection remain important information that should be combined with the molecular value to predict the net genetic merit of the individual candidates for selection. Phenotypic and molecular information can be combined in a single value called "selection index."

There are several ways of calculating selection indices using either phenotype data and/or molecular markers information. *Rindsel* is an R package that calculates both types of selection indices. *Rindsel* calculates: the Smith selection index [1], the Eigen selection index method (ESIM) [2, 3], the Restrictive Kempthorne

Delphine Fleury and Ryan Whitford (eds.), *Crop Breeding: Methods and Protocols*, Methods in Molecular Biology, vol. 1145, DOI 10.1007/978-1-4939-0446-4_8, © Springer Science+Business Media New York 2014

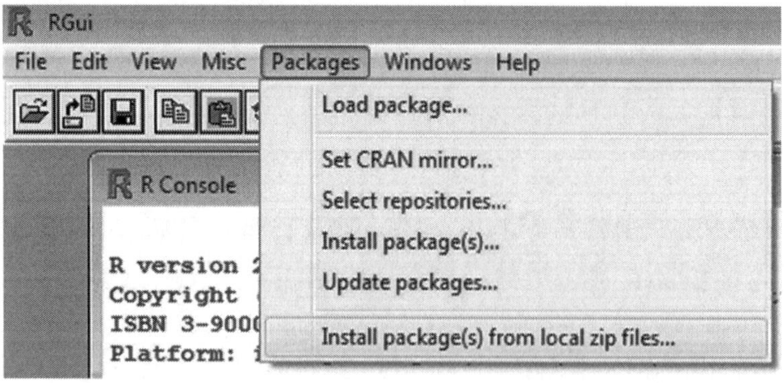

Fig. 1 R console showing the installation of Rindsel package

and Nordskog selection index [4], and the Restrictive Eigen selection index method (RESIM) [5]. These indices are defined as linear combinations of the observed mean phenotypic values of the traits of interest with the traits economic weights previously defined by the breeder. *Rindsel* also calculates molecular selection indices using Lande and Thompson method [6] and Molecular ESIM method (MESIM) [3]. A molecular selection index uses molecular marker or QTL information, i.e., the estimated effect of the QTL linked to the molecular marker (MQTL effect), that is, a combination of the MQTL effects and phenotypic information. The method aims to improve one trait by incorporating information on MQTL effects by means of the molecular selection indices. The final output incorporates all MQTL effects in one index.

2 Software Installation and Help

2.1 Installation

1. Install R from The Comprehensive R Archive Network (CRAN): http://cran.r-project.org/.
2. Save the R zip files of lme4 and Hmisc packages from CRAN.
3. Save the Rindsel zip file from the Integrated Breeding Platform tools repository https://www.integratedbreeding.net/.
4. Install the R packages lme4, Hmisc and Rindsel:

   From the menu, select Packages, then select Install package(s) from local zip file (Fig. 1). Select the zip file from the directory where you saved it.

2.2 Help

1. From the menu, select Help, then Html help (Fig.2). It will open the help browser.
2. Select the link packages and search for *Rindsel.*

   An alternative is to type help.search("*Rindsel*") in the R command prompt.

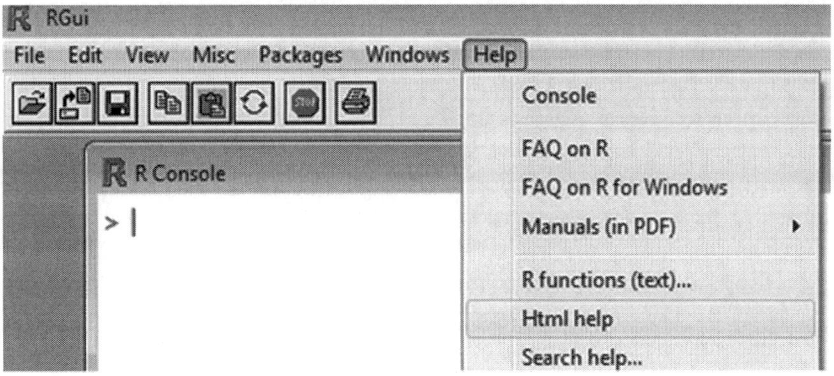

Fig. 2 R console showing how to call the Help menu

	A	B	C	D	E
1	REP	Block	ENTRY	MFL1	FFL1
2	1	12	1	103	101
3	1	12	2	106.86	106.67
4	1	24	3	97.75	101.67
5	1	10	4	103.33	105.14
6	1	19	5	95.3	99.33
7	1	22	6	102.25	104.33
8	1	19	8	103.25	104
9	1	14	9	100	100.6

Fig. 3 Organization of phenotypic data in an input Excel file (example in Rindsel\data\traits.csv)

3 Run *Rindsel*

3.1 Prepare the Input Files

1. Prepare the QTL file.
   Phenotypic data include field design and the trait data per entry (lines or plants). Create and Excel sheet. Data are organized as in Figure with replicates in first column, block in 2d column, entry number in the 3d column; then each trait is organized in the next columns (Fig. 3). The column headings are in the first line.
2. Prepare the markers file.
   Create an Excel spreadsheet. List the traits names in the first column, the indicator variable of the selected traits (0 absence, 1 presence) in the 2d column named Ind, the economic weights (−1 negative, 1 positive, necessary for Lande and Thompson index) in 3d column named eco, and the desired effect of selection (MESIMIndex) in the fourth column named sgn (−1 negative, 1 positive) (Fig. 4).

	A	B	C	D
1	TRAIT	Ind	eco	sgn
2	MFL1	1	-1	-1
3	FFL1	1	-1	-1
4	EHT1	1	-1	-1
5	PHT1	1	-1	-1
6	GY1	1	1	1
7	MFL2	0	-1	-1
8	FFL2	0	-1	-1
9	EHT2	0	-1	-1
10	PHT2	0	-1	-1
11	GY2	0	1	1
12	MFL3	0	-1	-1
13	FFL3	0	-1	-1
14	EHT3	0	-1	-1
15	PHT3	0	-1	-1
16	GY3	0	1	1

Fig. 4 Organization of weight data in an input Excel file (example in Rindsel\data\weights.csv). If known, heritability values can be used as traits' economic weights. In this example, the breeder wants to select plants of early flowering time in environment 1, which is defined by low values of male MFL1 and female FML1 plants. The desired effect for both traits is then set to −1. By contrast, lines of high grain yield in environment1 (GY1) will be selected, therefore the desired effect of GY1 is set to 1

	A	B	C	D
1	V	M1	M2	M3
2	1	1	1	1
3	2	-1	0	1
4	3	-1	0	0
5	4	1	0	1
6	5	0	1	0
7	6	0	-1	0
8	7	-1	-1	0
9	8	-1	-1	0

Fig. 5 Organization of markers data in the input Excel file (example in Rindsel\data\markers.csv). Markers names are headed in the first line

3. Prepare the weight file.
   Create an Excel spreadsheet. Alleles will be listed for each marker in columns. The alleles are noted 1 for AA, 0 for Aa, and −1 for aa (Fig. 5).
4. Prepare the phenotypic data file:
   Create an Excel spreadsheet. For each trait, "Score" is the QTL effect of the marker and reported in the odd columns (Fig. 6). Corresponding marker numbers are reported in the even column headed "Marker." Markers names are headed in the first line.

	A	B	C	D	E	F
1	Score	Marker	Score	Marker	Score	Marker
2	MFL1	MFL1	FFL1	FFL1	EHT1	EHT1
3	7.28	1	-0.42	1	2.65	11
4	-0.63	14	-0.84	10	1.36	14
5	6.65	2	0.34	11	-3.2	8
6	6.71	4	-1.49	13	2.39	9
7	0.11	4	0.48	17	2.82	4
8	-9.47	4	-0.59	3	-8.25	4
9	-6.66	5	0.53	4	-0.84	5
10	-1.09	6	-1.33	6	-8.18	6
11	1.53	7	-0.13	7	0.19	7

Fig. 6 Organization of QTL data in the input Excel file (example in Rindsel\data\qtl.csv)

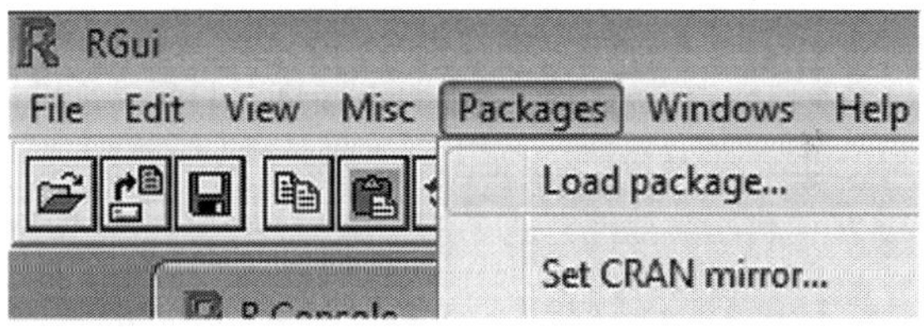

Fig. 7 R console showing how to load the Rindsel package

3.2 Load Rindsel

1. From the Packages menu select Load package (Fig. 7).
2. The available packages are displayed (Fig. 8). Select *Rindsel.*

3.3 Calculate Lande and Thompson Index

1. On the command prompt, write IndexName() to display the main menu (Fig. 9). Then type the index number you would like to calculate (in the example, 5 for Lande and Thompson index) and press Enter.
2. A window automatically opens requesting the phenotypic data file (Fig. 10). Browse and select the file.
3. A new window opens requesting the weight data file (Fig. 11). Browse and select the file.
4. The R routine starts by calculating the genetic and phenotypic covariance matrices. When the calculation is done, a new window opens requesting the markers file (Fig. 12). Browse and select the marker file.
5. A new window opens requesting the QTL data file (Fig. 13). Browse and select the file.
6. The output file is displayed in the R console (Fig. 14). Genetic, phenotypic, and molecular covariance matrices are displayed. A percentage of the individuals are automatically selected, by default 5 %, and sorted in LT index from highest to lowest.

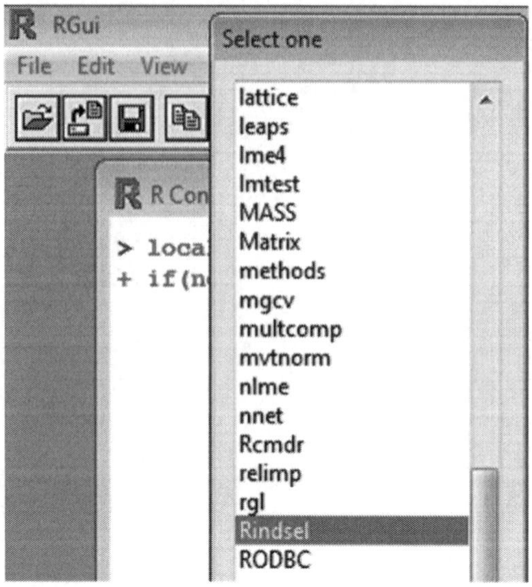

Fig. 8 R console showing how to start Rindsel

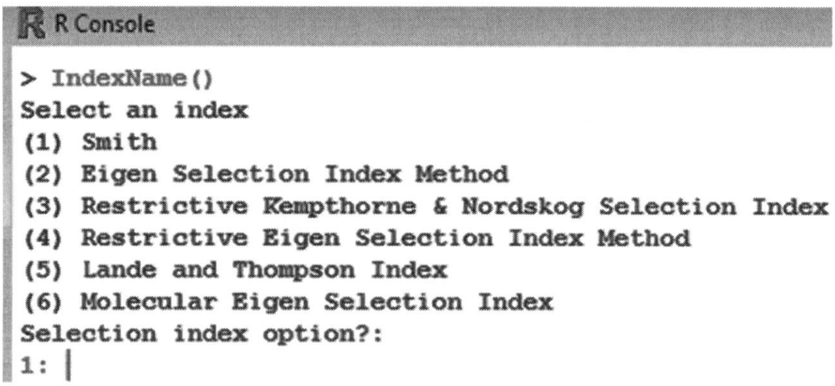

Fig. 9 R console showing how to run an index function

The values per trait and the LT index are also provided for all plants of the population.

3.4 Calculate Other Selection Index

1. For the other selection index methods, we proceed the same way by typing the number corresponding of the method, for example 6 for Molecular Eigen Selection Index.
2. It is possible to modify the default parameters for each method.

   For example, by typing MESIMIndex(selval = 10, rawdata = FALSE), selval = 10 means that we select the 10 % of traits with the highest values of the MESIM index, rawdat = FALSE means that we use covariance matrices that are already calculated. In the example above, we can use the covariance matrices obtained by calculating the LT index to calculate the MESIMIndex.

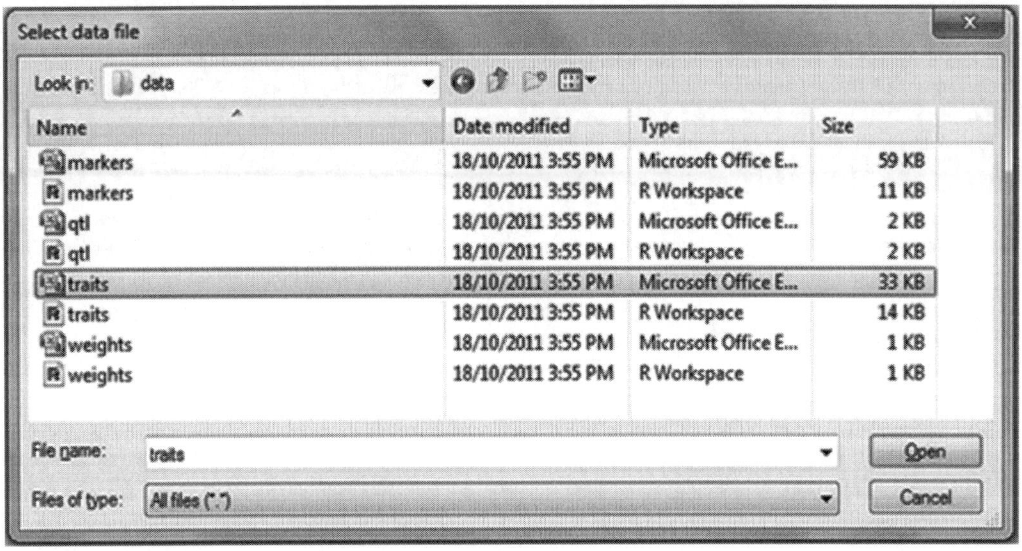

Fig. 10 R console showing how to load the phenotypic data file

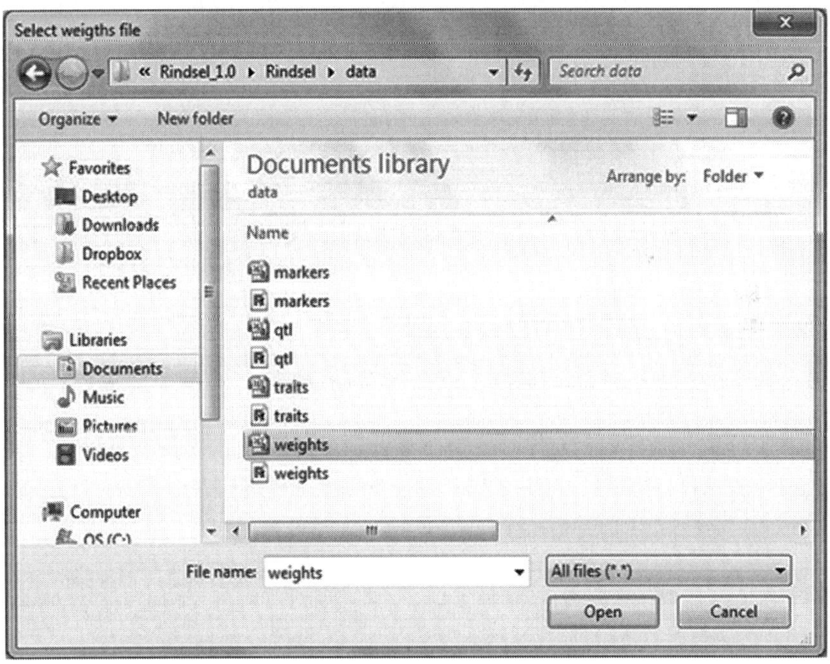

Fig. 11 R console showing how to load the weight data file

4 Notes

Simulation was used to compare the efficiency of selection indices (3). The simulation program used original data from actual doubled-haploid maize mapping population of 236 genotypes with five traits and their respective QTL. The five traits were male flowering

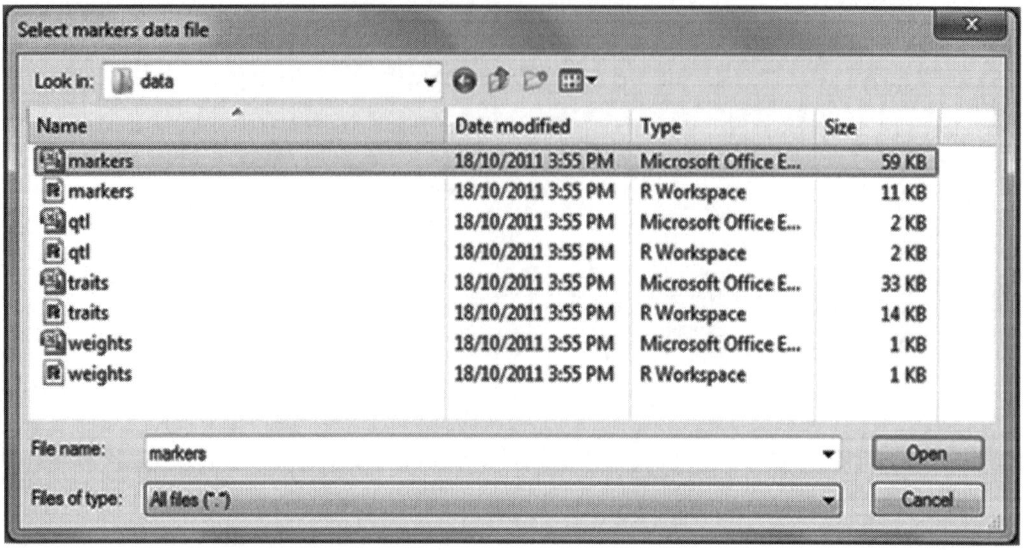

Fig. 12 R console showing how to load the marker data file

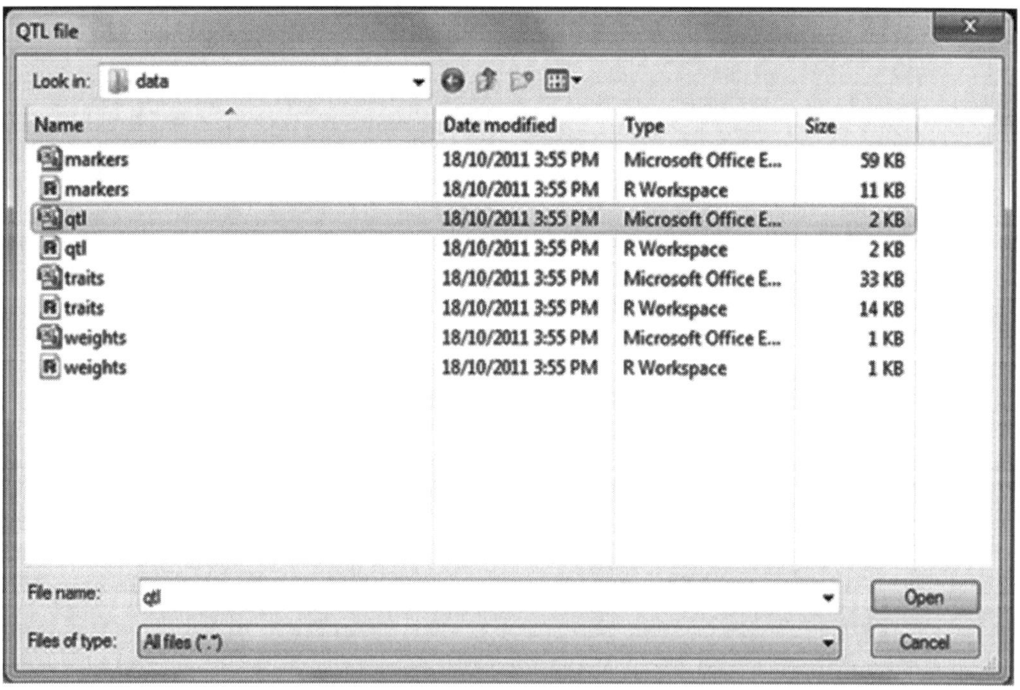

Fig. 13 R console showing how to load the QTL data file

time (MFL) (days), female flowering time (FFL) (days), plant height (PHT) (cm), ear height (EHT) (cm), and 100-kernel weight (HKF) (g). The data are provided with *Rindsel* package (under Rindsel/data folder) as an example dataset for training purposes.

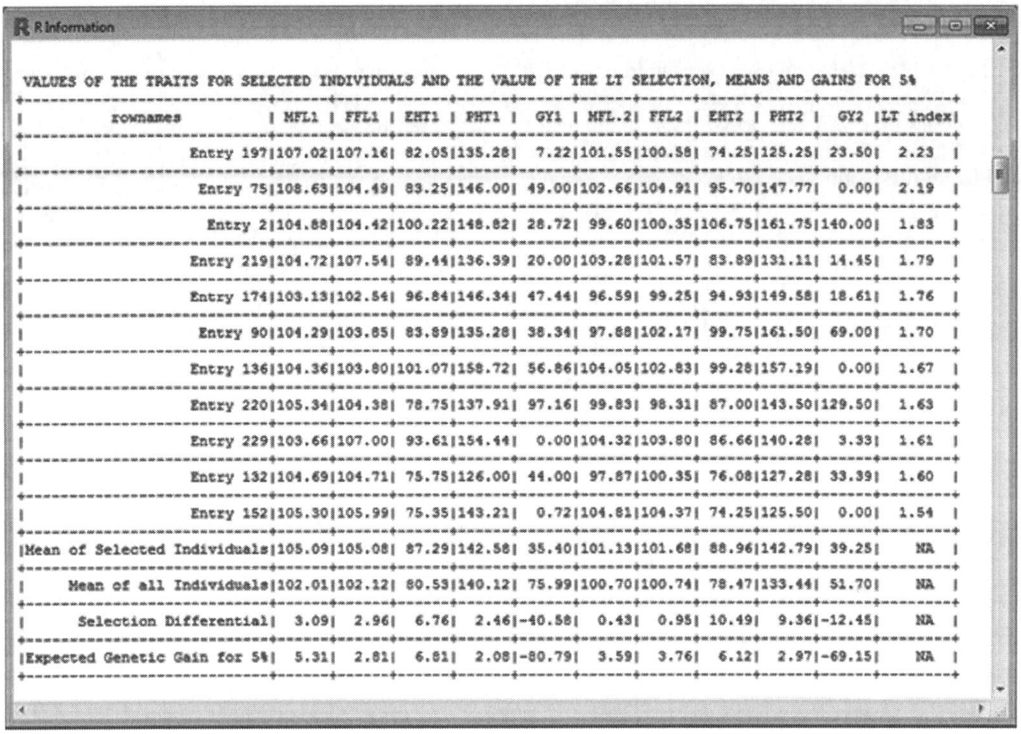

R Information

VALUES OF THE TRAITS FOR SELECTED INDIVIDUALS AND THE VALUE OF THE LT SELECTION, MEANS AND GAINS FOR 5%

rownames	MFL1	FFL1	EHT1	PHT1	GY1	MFL.2	FFL2	EHT2	PHT2	GY2	LT index
Entry 197	107.02	107.16	82.05	135.28	7.22	101.55	100.58	74.25	125.25	23.50	2.23
Entry 75	108.63	104.49	83.25	146.00	49.00	102.66	104.91	95.70	147.77	0.00	2.19
Entry 2	104.88	104.42	100.22	148.82	28.72	99.60	100.35	106.75	161.75	140.00	1.83
Entry 219	104.72	107.54	89.44	136.39	20.00	103.28	101.57	83.89	131.11	14.45	1.79
Entry 174	103.13	102.54	96.84	146.34	47.44	96.59	99.25	94.93	149.58	18.61	1.76
Entry 90	104.29	103.85	83.89	135.28	38.34	97.88	102.17	99.75	161.50	69.00	1.70
Entry 136	104.36	103.80	101.07	158.72	56.86	104.05	102.83	99.28	157.19	0.00	1.67
Entry 220	105.34	104.38	78.75	137.91	97.16	99.83	98.31	87.00	143.50	129.50	1.63
Entry 229	103.66	107.00	93.61	154.44	0.00	104.32	103.80	86.66	140.28	3.33	1.61
Entry 132	104.69	104.71	75.75	126.00	44.00	97.87	100.35	76.08	127.28	33.39	1.60
Entry 152	105.30	105.99	75.35	143.21	0.72	104.81	104.37	74.25	125.50	0.00	1.54
Mean of Selected Individuals	105.09	105.08	87.29	142.58	35.40	101.13	101.68	88.96	142.79	39.25	NA
Mean of all Individuals	102.01	102.12	80.53	140.12	75.99	100.70	100.74	78.47	133.44	51.70	NA
Selection Differential	3.09	2.96	6.76	2.46	-40.58	0.43	0.95	10.49	9.36	-12.45	NA
Expected Genetic Gain for 5%	5.31	2.81	6.81	2.08	-80.79	3.59	3.76	6.12	2.97	-69.15	NA

Fig. 14 Output results displayed in the R console showing the Lande and Thompson index per trait and per entry line (LT)

Simulation results show that when genotypes are selected on the basis of individual traits, MESIM increased the response to selection over the Lande and Thompson index. When several traits are selected simultaneously, MESIM outperformed Lande and Thompson for traits with low heritability. For traits with high heritability, ESIM performed very well. MESIM can be considered a generalization of ESIM (2) when information on QTL is incorporated through molecular markers. Lande and Thompson index requires economic weights, which can be chosen as −1, −1, −1, −1, and 1 for each trait in the maize example above. Heritability of the traits can also be used as weights. However it can be difficult to choose weights when no estimates of economic weights are available. MESIM has the advantage over Lande and Thompson to do not require economic weight.

Acknowledgments

This work is supported by the Generation Challenge Programme through the Integrated Breeding Platform project and CIMMYT.

References

1. Smith HF (1936) A discriminant function for plant selection. In: Papers on quantitative genetics and related topics. Department of Genetics, North Carolina State College, Raleigh, NC, pp 466–476
2. Cerón-Rojas JJ, Crossa J, Sahagún-Castellanos J, Castillo-González F, Santacruz-Varela A (2006) A selection index method based on Eigen analysis. Crop Sci 46:1711–1721
3. Cerón-Rojas J, Castillo-González F, Sahagún-Castellanos J, Santacruz-Varela A, Benítez-Riquelme I, Crossa J (2008) A molecular selection index method based on Eigen analysis. Genetics 180:547–557
4. Kempthorne O, Nordskog AW (1959) Restricted selection indices. Biometrics 15:10–19
5. Cerón-Rojas JJ, Sahagún-Castellanos J, Castillo-González F, Santacruz-Varela A, Crossa J (2008) A restricted selection index method based on Eigen analysis. J Agric Biol Environ Stat 13:421–438
6. Lande R, Thompson R (1990) Efficiency of marker-assisted selection in the improvement of quantitative traits. Genetics 124:743–756

Chapter 9

OptiMAS: A Decision Support Tool to Conduct Marker-Assisted Selection Programs

Fabio Valente, Franck Gauthier, Nicolas Bardol, Guylaine Blanc, Johann Joets, Alain Charcosset, and Laurence Moreau

Abstract

Ongoing major advances in plant genotyping and phenotyping lead to a better understanding of genetic architecture of agronomical traits. In this context, it is important to develop decision support tools to help breeders in implementing marker-assisted selection (MAS) projects to assemble new allele combinations. Algorithms have been developed within an interactive graphical interface to (a) trace parental QTL alleles throughout selection generations, (b) propose strategies to select the best plants based on estimated molecular scores, and (c) efficiently intermate them depending on the expected value of their progenies. By investigating multi-allelic context and diverse pedigree structure, OptiMAS enables to assemble favorable alleles issued from diverse parents and further accelerate genetic gain.

Key words Marker-assisted selection, Plant breeding, QTL, Multiparental designs, Gene pyramiding

1 Introduction

Molecular markers have been extensively exploited in crop breeding since the 1980s, to evaluate diversity within breeding populations, saturate genetic maps, and detect markers linked to quantitative trait loci (QTL) or genes controlling traits of economical importance [1]. For simple traits, these markers can be applied for diagnostic purposes in early screening process (e.g., selection against disease susceptibility) or targeted replacement of chromosomal segments by means of marker-assisted backcrossing. For Quantitative traits, Lande and Thompson [2] advocated using markers significantly associated with QTL to predict the genetic value of individuals (molecular score), select the best ones, and conduct marker-assisted selection (MAS) cycles in order to assemble favorable alleles. Many studies theoretically investigated strategies of MAS that were applied with success, particularly in private companies [3].

Delphine Fleury and Ryan Whitford (eds.), *Crop Breeding: Methods and Protocols*, Methods in Molecular Biology, vol. 1145, DOI 10.1007/978-1-4939-0446-4_9, © Springer Science+Business Media Newyork 2014

However, these were mainly conducted in the context of biparental populations. By addressing a broader diversity, multiparental connected designs (1) increase the power and the accuracy of QTL detection, and (2) enable to estimate simultaneously the different parental allele effects and identify the most favorable ones for selection [4–7]. Such experiments can be exploited to create a given ideal genotype (ideotype) assembling alleles from diverse parental origins.

OptiMAS [8] has been developed in the framework of Integrated Breeding Platform (IBP, https://www.integrated-breeding.net) to provide help to plant breeders for making rapid and efficient selection decision through a user-friendly interface, considering the general framework of multiparental designs and complex pedigree structure generally observed in applied crop breeding programs. To do so, OptiMAS main algorithms have been deployed to trace parental QTL alleles identified as favorable throughout selection generations, using information given by markers located in the vicinity of the estimated QTL positions. Probabilities of allele transmission are computed in different MAS schemes and mating designs (intercrossing, selfing, backcrossing, double haploids, recombinant inbred lines) with the possibility of considering generations without genotypic information. Then, strategies are proposed to select the best plants based on estimated molecular scores (MS) and to efficiently intermate them depending on the expected value of their progenies.

2 Materials

OptiMAS is available in two versions. The first one (not presented in this chapter) manages computationally intensive processes for step 1. It is written in C-ANSI language and runs in command line, which provides convenient integration possibilities within custom analysis pipelines and databases. The second version integrates the C program within a graphical user interface (GUI) coded in C++ using Qt, Qwt, and Graphviz libraries which provides additional breeding oriented functionalities. Installable versions are distributed to run under most modern GNU/Linux, Windows (XP/7), and Mac OS X (10.5 or later with Intel processor) systems. Ready-to-use binaries installation packages and the source code, which you are welcome to attempt to compile on your favorite platform, are available and can be downloaded at http://moulon.inra.fr/optimas. OptiMAS is provided with two example datasets corresponding to bi- and multiparental programs, and an integrated manual accessible through a help function.

3 Methods

3.1 Case Study Description

Functionalities of OptiMAS will be presented here based on real data coming from a multiparental marker-assisted recurrent selection (MARS) study [5, 6]. It corresponds to one of the two examples provided with the software, so that operations described below can be reproduced by the user. Six connected F2 populations, with 150 individuals each, were obtained from a half-diallel design between four unrelated maize inbred lines (DE, F283, F810, and F9005). Eleven QTL were detected for silking date and the objective of the project is to assemble the earliest alleles at all these QTL. To do so, a set of 34 markers was selected with at least three markers (microsatellites) to follow each QTL. Two cycles of MARS were performed with each time a step of selfing before intermating. In this example, OptiMAS is used at the last cycle to select the best individuals (among 297 genotyped plants) that will be used for the next cycle of MARS (Fig. 1).

3.2 Data Preparation

To run the program, two input files (genotypes/pedigree file and the map file) containing all information about this MAS design must be supplied (*see* **Note 1**).

3.2.1 Genotypes/Pedigree File (.dat)

The genotypes/pedigree file contains individuals, pedigree information, and genotypic data. The input file organization corresponding to the MAS pedigree in Fig. 1 is exemplified below (Table 1, *see* **Note 1**).

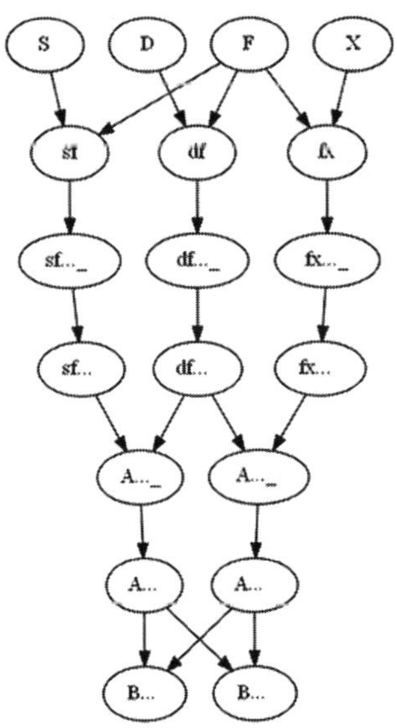

The four inbred lines, D (*DE*), F (*F283*), S (*F810*) and X (*F9005*) corresponding to the parental alleles (*d*, *f*, *s* and *x*, respectively) to follow through generations of selection.

Three out of six F1 plants (F2 progenies of sx, dx, sd F1 were not selected at the F2 cycle).

F1 hybrids have been selfed to obtain three F2 populations of 150 genotyped individuals. 25 F2 candidate plants were selected (first cycle of selection).

Two other selfing operations have been done. 25 families of F4 individuals were produced (without genotyping information) and crossed.

Among the progenies, 21 candidates were selected based on genotyping information (second cycle named "C1").

21 selfed families were produced (without genotyping information) and crossed together.

297 individuals issued from these crosses were genotyped (third cycle of selection named "C2". OptiMAS is used at this step to select the best individuals that will be used for the next cycle of MARS.

Fig. 1 Multiparental MAS study used as example within OptiMAS [5, 6]

Table 1
The genotypes/pedigree file for a sample of individuals with markers used to follow QTL1 and QTL2 (full table extends on the right to markers associated with QTL 11)

Id	P1	P2	Step	Cycle	Group	Pr443	Pr256	Pr299	Pr454	Pr1204
D	d	d	IL			1	3	3	2	2
F	f	f	IL			1	3	2	1	3
S	s	s	IL			1	1	1	1	1
X	x	x	IL			2	2	2	2	2
df	D	F	CR							
fx	F	X	CR							
sf	S	F	CR							
df108_	df	df	S1	F2		1	3	2/3	1/2	2/-
fx305_	fx	fx	S1	F2		2	2	2	1	3
sf648_	sf	sf	S1	F2		1	3	2	1	1/-
sf650_	sf	sf	S1	F2		1	1	1/2	1	1/-
df108	df108_	df108_	S2							
fx305	fx305_	fx305_	S2							
sf648	sf648_	sf648_	S2							
sf650	sf650_	sf650_	S2							
A23_	df108	sf648	CR	C1		1	3	2/3	1/2	-
A167_	fx305	df108	CR	C1		1/2	2/3	2/3	1/2	2/3
A1005_	df108	sf650	CR	C1		1	1/3	2	1	-
A23	A23_	A23_	S1							
A167	A167_	A167_	S1							
A1005	A1005_	A1005_	S1							
B124	A23	A167	CR	C2		1/2	2/3	2	1/2	3
B8	A1005	A1005	CR	C2		1	1	2	1	1/3

1. The "Id" column names each individual coded as a character string without any blanks and special characters. It must be unique.
2. "P1, P2" correspond to the name of the parents of the individual (must exist as individuals above in the file except for founder parents). Individuals must be ranked according to generations (from oldest to most recent). Founder parents of the program (e.g., D, S, F, and X) are assumed to be homozygous

Table 2
Genetic map with selected markers used to follow QTL1 and QTL2 (full table extends to the bottom down to QTL 11 and associated markers)

Locus	Chr	QTL	Pos	All+
Pr443	1	1	0	
qtl1	1	1	23.3	x
Pr256	1	1	28.5	
Pr299	1	2	0	
qtl2	1	2	8	f
Pr454	1	2	14.5	
Pr1204	1	2	36.3	

lines with no residual heterozygosity. Their pedigree is assumed to be unknown. "P1, P2" columns indicate in this case the allele that will be transmitted through generations. This allele is identified as a single character (e.g., d, s, f, and x).

3. The "Step" column denotes the pedigree relationship between the individual and its parents. "CR" (Cross) indicates that the individual results from a cross between its two parents; "S*n*," means that the individual results from *n* (integer number between 1 and 20) generations of selfing of its parent (in this case the two parents Id must be identical); "RIL" for recombinant inbred lines, assumes that the individual results from an infinite number of selfing generations from an initial F1 hybrid; "DH" for double haploids, assumes that the individual results from haplo-diploidization from an initial F1 hybrid; "IL" indicates founder inbred status.
4. "Cycle/Group" columns show optional information regarding the generation in the program (e.g., first cycle, second cycle, F2, and F4) or another classification criterion (e.g., subprograms, families). These two columns can be left blank (or "-").
5. Markers 1–*n* refer to marker names, as described in Table 2, and genotyping results. The software can deal with SNP, microsatellites and any bi/multi allelic marker genotyping technique with either dominant or codominant scoring (*see* **Note 2**).

The possibility of including non-genotyped individuals (e.g., fx305) enables to analyze most common MAS schemes and mating designs. So, if a specific individual derives from one or several generation(s) with non-genotyped ancestors, one must add these ancestors in the data file. Note that multiple selfing generations can be managed directly through the "S*n*" descriptors.

3.2.2 Genetic Map (.map)

The aim is to create a "target genotype" (ideotype) with all favorable alleles at the QTL positions. So, before running OptiMAS it is necessary to define the parental alleles to assemble. The map file (Table 2 and *see* **Note 3**) is supplied by the user to specify the information regarding markers, QTL and identification of favorable alleles, issued from a QTL mapping analysis (possibly considering several traits, environments, etc.).

The QTL position is rarely located at a marker; consequently QTL alleles are unknown and must be inferred from flanking markers. Thus, it is very important to select a subset of markers as informative as possible (especially in multi-parental context) to follow the favorable parental alleles (based on haplotypes).

1. The "Locus" column designates markers/QTL. The name of the QTL must start with "qtl" (e.g., qtl1, qtl_umc11, qtl_1 ...). Any other names are by default considered as markers. The only constraint on names is to avoid blank in character chain.
2. "Chr" refers to index (numerical value) of the chromosome where the marker/QTL is located.
3. The "QTL" column points affiliation of markers to specific QTL (e.g., two markers Pr443/Pr256 are used to follow allele transmission for QTL 1). Indexes for each QTL must be unique, continuous (no gap in list) starting from "1" along the genome.
4. "Pos" corresponds to the relative chromosomal positions of loci. For QTL, this is the estimated QTL position coming from the QTL detection results. Positions of the different loci must be obtained using the Haldane's mapping function (i.e., by assuming no interference).
5. The "All+" column indicates the parental allele(s) considered as being favorable (target genotype). For QTL 1, the favorable allele "x" refers to the parental line named "X" (*see* columns "P1"/"P2" of Table 1). For QTL 4 (not represented in Table 2), "s/f" refers to parental lines (S and F) which can be considered as favorable relatively to other parental lines.

3.3 Running OptiMAS Graphical User Interface (GUI)

1. On PC and Macintosh computers, click on the OptiMAS icon to start a session. On GNU/Linux systems, type "optimas_gui &."
2. The three different modules corresponding to the different steps of a selection program are represented by three icons on the left side (Fig. 3).
3. To run the program, select "File" menu, then "Import Data..." in the OptiMAS window.
4. The "Import Data/Settings" window will show up (Fig. 2). Load the two previously defined input files (*see* Subheadings 3.2.1 and 3.2.2) using the browser.

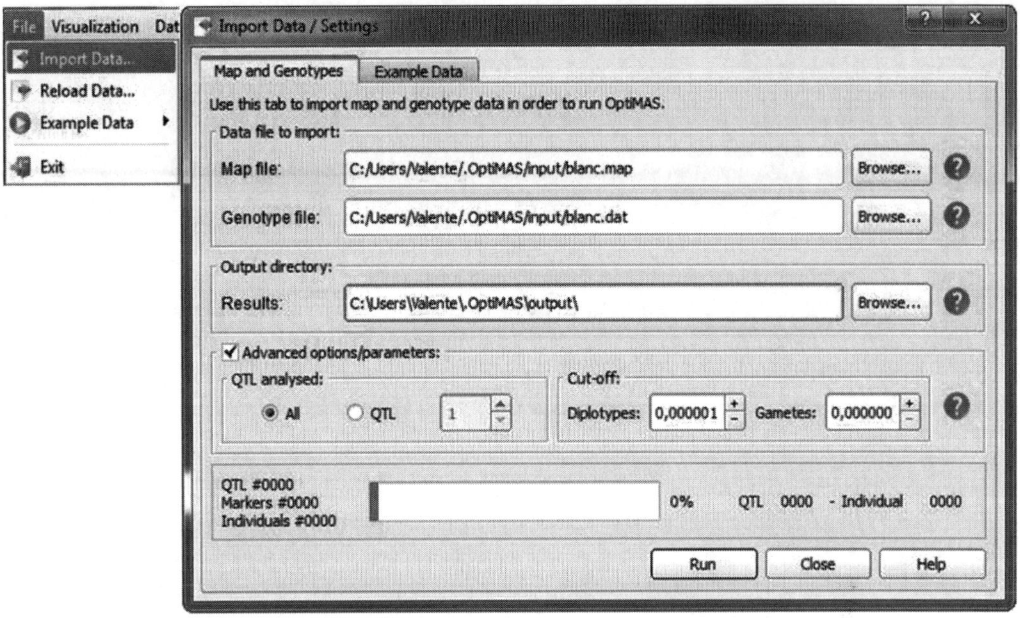

Fig. 2 Upload of data sets, definition of output directory and advanced options/parameters

5. Click on the top "browse…" button to load the map file. In the new window displaying the user's subdirectories and files, select a file containing the genetic map.
6. Do the same thing for the genotypes/pedigree file.
7. Click on the third "browse…" button to select the output directory (path to the folder where the results will be stored). Results from each run will be stored within a new dated directory created automatically within this folder. The output directory should not be in the "Program Files" folder or other specific directories with administrator privileges.
8. Default options will be used in this example but other parameters/options can be selected by checking the "Advanced options/parameters" button (*see* **Note 4**).
9. Click on the "Run" button to analyze the data set. The program will create output results in the folder that you specified.
10. At the end of the run (the progress bar displays 100 %), close the "Import data…" window by pressing the "Close" button. If a warning message appears, report to **Note 5** before any interpretation.

To visualize and analyze the results, the OptiMAS GUI includes three different modules corresponding to the different steps of a selection program.

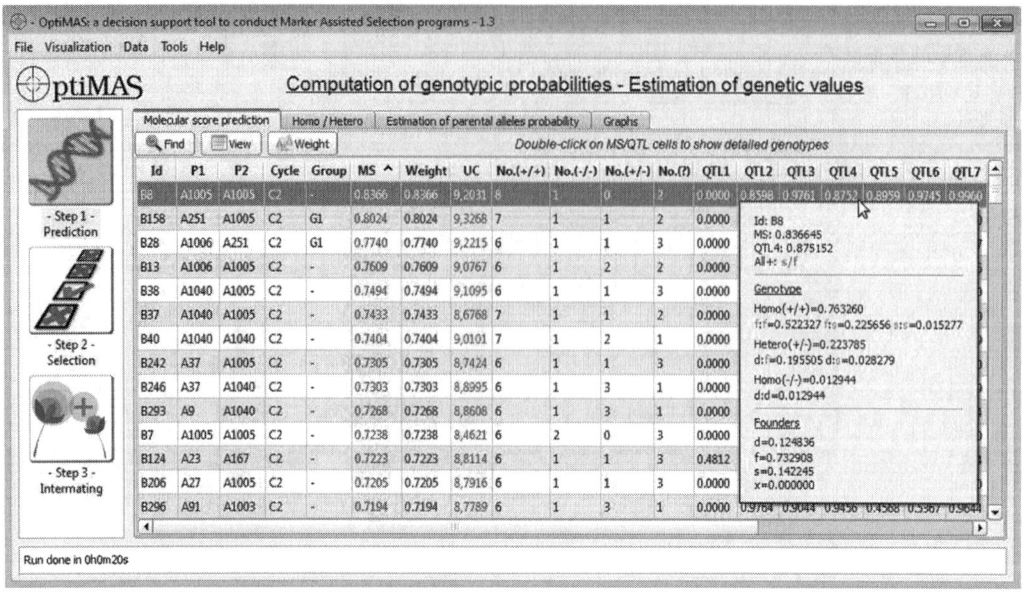

Fig. 3 OptiMAS main table for step 1. This table displays prediction of global scores and genotypes at QTL (with detailed view for one cell)

3.4 Step 1: Computation of Genotypic Probabilities: Estimation of Genetic Values

At the end of the run, results are displayed via tables and graphs.

1. In the main table ("Molecular score prediction", Fig. 3) individuals are represented in lines followed by their pedigree, cycle of selection, group they belong to, synthetic indices as described below and a molecular score at each QTL (e.g., QTL4 = 0.8752). This molecular score corresponds to the expected probability of favorable allele(s) at the QTL, based on the classification of parental alleles into favorable and unfavorable categories. The MS column refers to the average molecular score across all QTL and varies between 0 and 1. A value of 1 indicates that the individual corresponds to the targeted ideotype (it is homozygous for the favorable allele(s) for this QTL or all the QTL).
2. Clicking on the MS or a QTL column sorts plants according to their molecular score or QTL score.
3. Double clicking on a QTL cell shows detailed genotype probabilities for this specific cell (*see* "Genotype" section of Fig. 3). This view summarizes and aggregates the information presented for all individuals in the two other tables ("Homo/Hetero" and "Estimation of parental allele probabilities"), accessible by clicking on the corresponding tab. For each candidate individual, the probabilities to be homozygous or heterozygous at each QTL are computed according to favorable/unfavorable grouping of founder alleles. For example at QTL4, individual B8 has a probability of 0.763260 to be

homozygous for the favorable alleles s/f. This score (Homo(+/+) = 0.763260) corresponds to the sum of the probabilities of the genotypes f:f = 0.522327, s:f = 0.225656, and s:s = 0.015277. Founders (represented by d, f, s, and x alleles) indicate the expected proportion of parental alleles at QTL positions (e.g., at QTL4 of individual B8, d = 0.124836, f = 0.732908, s = 0.142245, and x = 0.0).

4. Pressing the "Weight" button opens a dialog window in order to give more or less importance to the different QTL (weighted average of QTL molecular score values). It is possible to discard QTL and/or to attribute economical weights defined by the breeder in order to compute a "Weight" index. Assigning a weight of 3.00 at QTL1 and pressing the "Apply" button updates the "Weight" column and therefore produces a new ranking of individuals.
5. Clicking on the "UC" column sorts individuals according to a third criterion, "Utility Criterion," which evaluates the expected value of superior gametes of each individual by combining the molecular score with the expected variance of the MS of its gametes (*see* **Note 6**).
6. Selecting the "Visualization" menu, then the "Color scheme" option opens a new window displaying a set of parameters (cut-off and colors). A value of 0.75 (by default) is selected for the probability threshold to be considered as homozygous (un)favorable or heterozygous at the QTL positions. A color can be assigned to each of them. Genotypes which are not assigned to any of these categories are considered "uncertain genotypes (?)." When a new set of parameters (cut-off/colors) is applied, the four corresponding columns (No.(+/+), No. (−/−); No(+/−): number of QTL homozygous for (un)favorable allele(s) or heterozygous; No.(?): number of QTL defined as uncertain) of the "Molecular score prediction" table are updated and a new colored view of the molecular score table is displayed to identify more easily QTL for which a given individual is considered as fixed or not (Fig. 4 and *see* **Note 7**).
7. Clicking on the "Graphs" tab displays the distribution of several indicators (QTL molecular scores, MS) and their evolution over the different cycles of selection.

3.5 Step 2: Selection of Individuals

Taking into account all the information coming from the previous tables, individuals can be selected for producing the next generation using three different ways.

3.5.1 Manual Selection

Individuals are selected manually by the user in the "Step 1" window ("Molecular score prediction" table) based on his/her own evaluation.

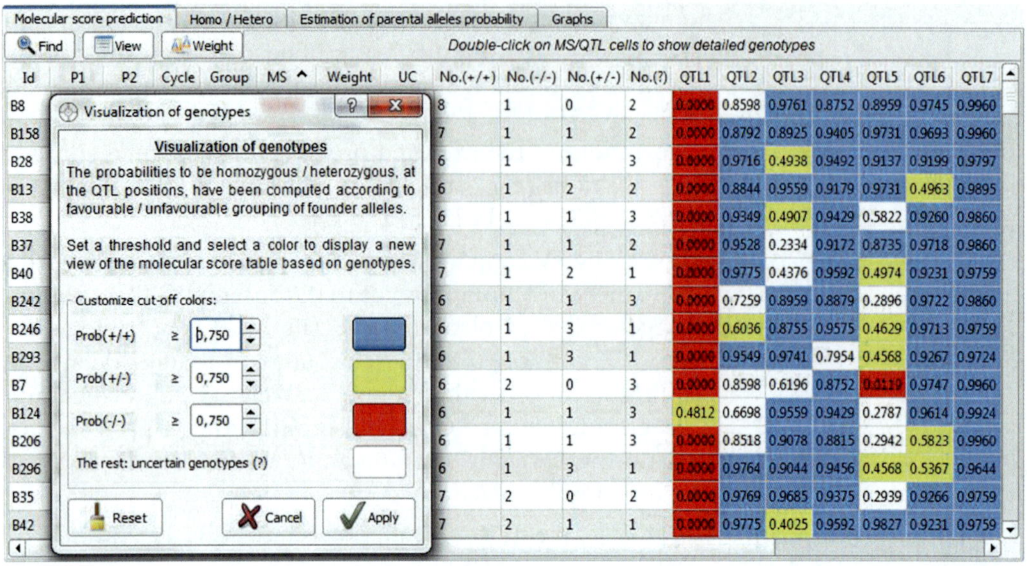

Fig. 4 Colored view of the molecular score prediction table. *Left panel* displays window for thresholds and colors definition

1. Plants can be selected via click and drag selection (or simple click + Ctrl). For example, select individuals B124 and B125 (the "Find" button can be used to locate these two individuals).
2. Press right click, click "Add to list...," select your list (e.g., "new_list" can be renamed in "List1_Manual_Selection" by double-click).
3. Press "Ok" to add these selected individuals to the list of your choice.
4. Click on the "Step 2" icon to visualize and access to this new list. Names of lists can be modified at all steps (by double clicking on the name of the list one wants to modify).

3.5.2 Truncation Selection (TS)

Individuals can be ranked automatically based on the three possible criteria: Molecular Score (MS), Weighted MS (Weight), or Utility Criterion (UC). The N_{sel} first sorted individuals are selected to generate the next population.

1. To select the ten first individuals based on the molecular score, go to the "Step 2" page and set up $N_{sel} = 10$, Criterion = Molecular Score.
2. Add them to a new list named "List2_Truncation_MS_Selection." Double-click on the empty list named "List_Selection2" and rename it in "List2_Truncation_MS_Selection." Select this new empty list (i.e., List = List2_Truncation_MS_Selection).
3. To constrain these ten individuals to be selected in the last cycle, click on the "Option..." button and select Cycle = C2.

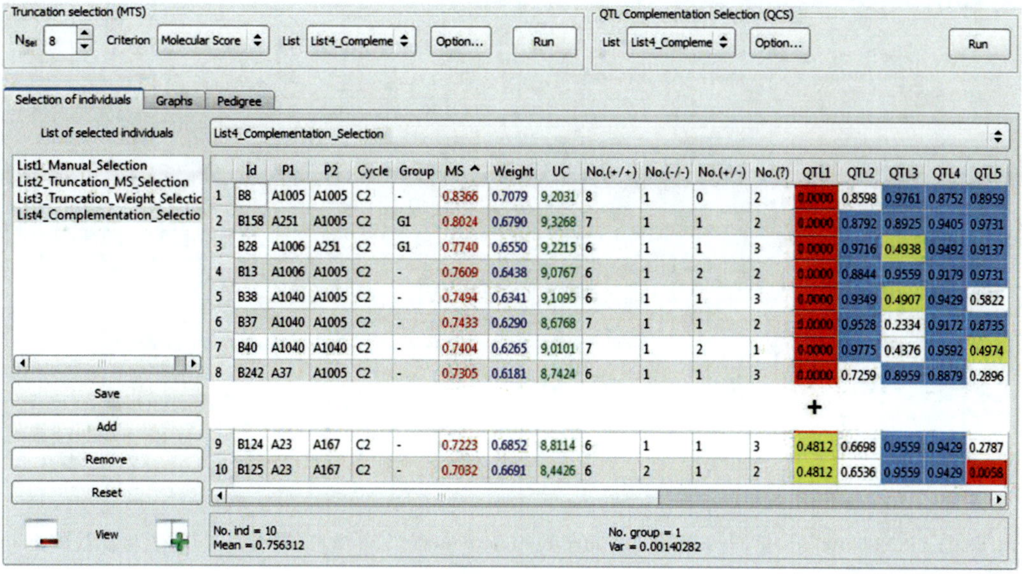

Fig. 5 Outcome of QTL complementation selection method. Individuals 9 and 10 have been added to the initial list of 8 to keep favorable allele at QTL1

4. Press "Run" to see the ten first individuals, based on the molecular score, added to the new created list.
5. Follow the same steps to create a third list with the ten first individuals based on the weighted molecular score (i.e., Press the "Add" button to create a new empty list, then select Criterion = Weight and List = List3_Truncation_Weight_Selection).

3.5.3 QTL Complementation Selection (QCS)

The QCS takes into account complementarities between candidate individuals regarding the favorable alleles they carry. It aims at preventing the loss of rare favorable alleles [9]. This option is important when a high number of QTL is considered.

1. Apply the truncation selection method (*see* Subheading 3.5.2) to create a fourth list named "List4_Complementation_Selection" with the eight first individuals based on the molecular score. The eight first individuals (B8… B242, Fig. 5) are added to the list. The colored view illustrates that we are losing the favorable allele for QTL 1 (in red, QTL1 = 0 for all individuals). To overcome this issue, the QCS can be applied in order to find two additional individuals such that their QTL composition complements those of the eight individuals already selected (Fig. 5).
2. Select the "List4_Complementation_Selection" in the QCS section.
3. Click on the QCS "Option…" button to set up the parameters shown in Fig. 6 (on the left side).

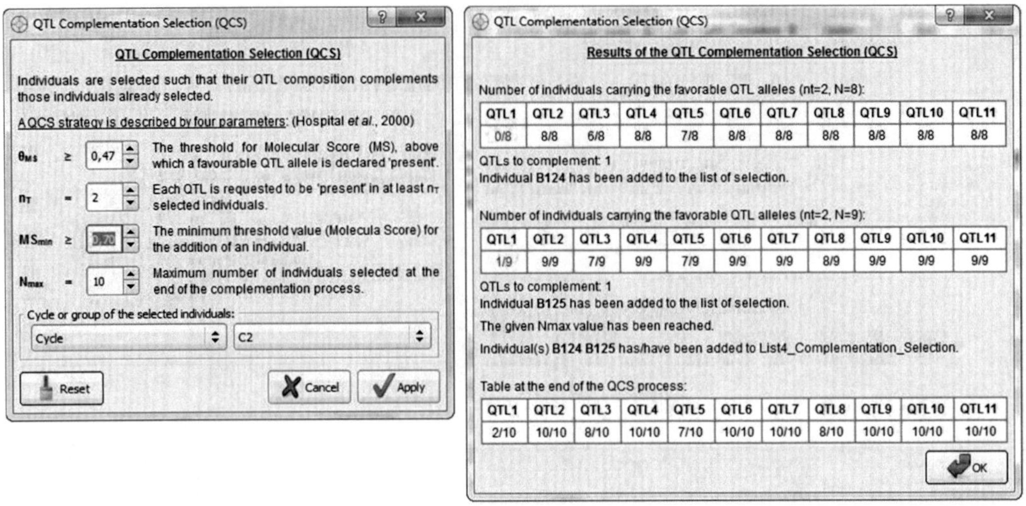

Fig. 6 QTL complementation selection outcome (on the *right*) according to parameters defined in QCS settings window (on the *left*)

4. The first parameter θ_{MS}, corresponds to the MS QTL threshold (QTLx > 0.47 kept by default), above which a favorable QTL allele is declared "present". In this case, not only individuals considered as homozygous for the favorable allele at QTL position will be considered but also heterozygous individuals (Fig. 5, in yellow).
5. The second parameter, n_T, sets that each favorable QTL allele is requested to be "present" in at least n_T selected individuals (here $n_T = 2$).
6. Set a minimum MS threshold value (here $MS_{min} \geq 0.70$) for the addition of an individual (i.e., individuals with MS (genetic value) < 0.7 are not considered).
7. The N_{max} parameter is the maximum number of individuals selected at the end of the QCS process. Here, up to two individuals will be added to the final subset of selected individual (i.e., $N_{max} - 10$).
8. By selecting Cycle = C2 instead of "None" by default, only individuals that belong to the last cycle of selection (C2) can be selected.
9. Press "Apply" and then "Run." The result will appear in a pop-up window (Fig. 6, on the right). With the present set of parameters, two individuals (B124, B125) have been added to the selected list. *See* **Note 8** for more details about the algorithm used in the QCS process.

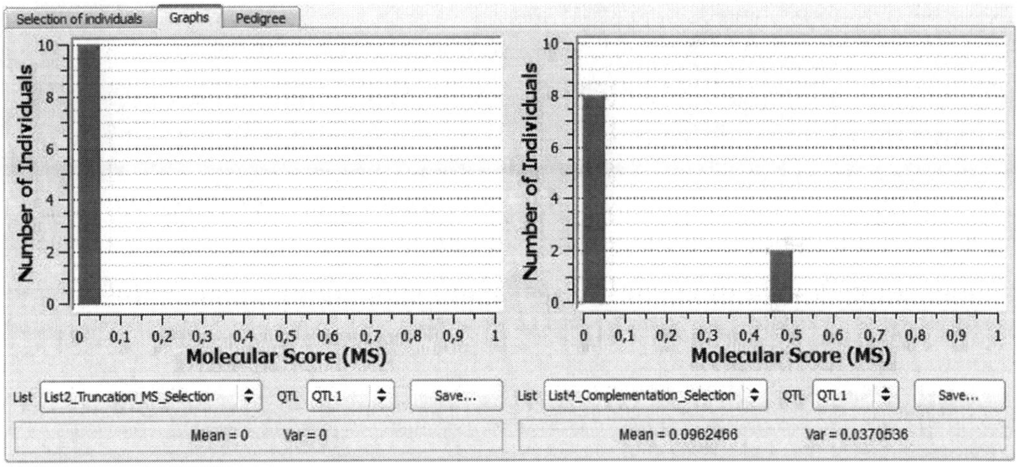

Fig. 7 Comparison between lists of selected individuals. Graphs display distribution of molecular scores (here for QTL1)

3.5.4 Display and Comparison of Lists of Selected Individuals

The different lists of selected individuals can be compared in two parallel tables, graphs (Fig. 7) and via their pedigree (Fig. 8).

1. Click on the "Selection of individuals" tab. Select the "List2_Truncation_MS_Selection" on the left table and the "List4_Complementation_Selection" on the right. The two last individuals (B246 and B293) are replaced by B124 and B125 in the QCS list.
2. Click on the "Graphs" tab. Select the truncation selection list on the left graph (i.e., List = List2_Truncation_MS_Selection, QTL = QTL1) and the QCS list on the on the right side (i.e., List = List4_Complementation_Selection, QTL = QTL1). These two graphs display the distribution of selected individuals according to their molecular score at QTL1. It shows that using truncation selection based on MS leads to the selection of individuals that all carry the unfavorable allele at QTL1 (Fig. 7, histogram on the left), whereas the two individuals added via the QCS procedure are likely heterozygous (MS close to 0.5, histogram on the right).
3. Click on the "Pedigree" tab. Select the truncation selection list (i.e., List = List2_Truncation_MS_Selection), select the "Alone" option (to only have the individuals of the selected list), and click on "Generate." Save the picture in the format of your choice via the "Save..." button. Follow the same procedure for the QCS list in order to display the pedigree, visualize, and compare the origin of the selected individuals of each list. The two individuals added via the QCS bring the parental allele x at QTL1 and therefore reintroduce parent X in the pedigree (Fig. 8; *see* **Note 9**).

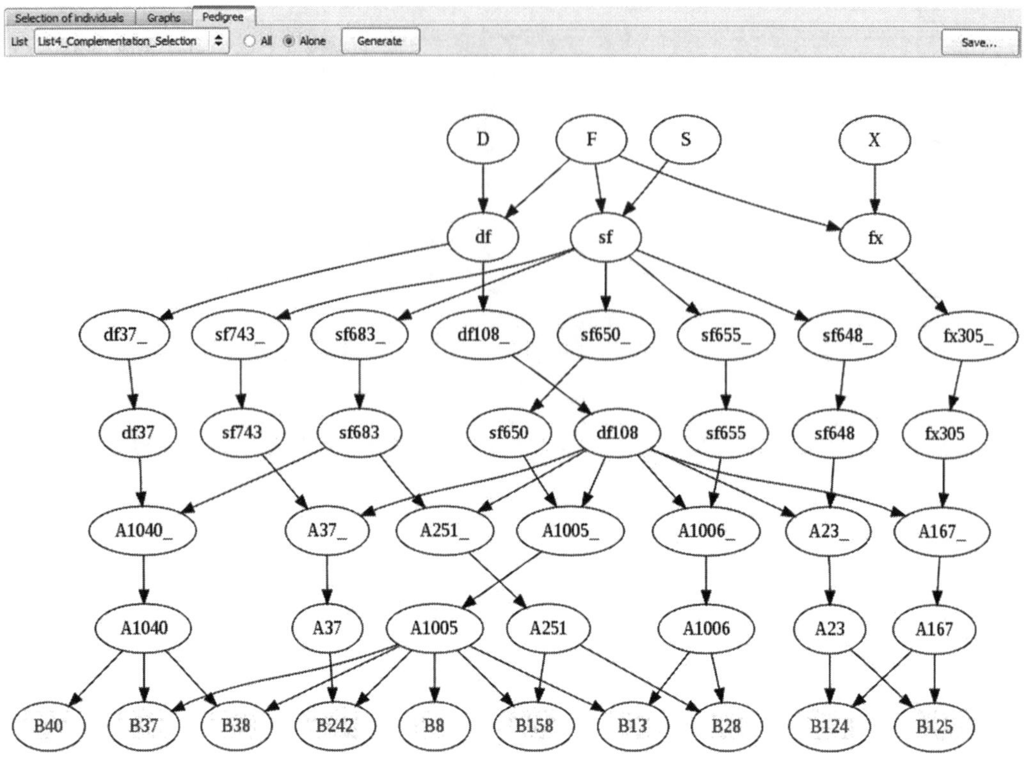

Fig. 8 Display of pedigree of the individuals selected via the QCS method

3.6 Step 3: Identification of Crosses to Be Made Among Selected Individuals

After selecting the individuals, we must identify the crosses to be made for the next MARS cycle. Crosses can be made between individuals of a single list (diallel design) or two complementary lists (factorial design). The diallel situation can be managed with three options: (1) the automatic definition of the whole list of possible crosses according to a half-diallel (complete method), (2) the "better-half" strategy [10] which consists of avoiding crosses between selected individuals with the lowest scores, and (3) application of constraints on the contribution of parents or on the maximum number of crosses. In this last case, best crosses are determined according to either the (weighted) molecular score or the utility criterion. Then, lists of crosses created via the different methods can be analyzed and compared via graphs.

3.6.1 Half-Diallel and Better-Half Designs

1. Move to the next step by selecting the "Step 3 – Intermating" icon.
2. Rename the first empty list of crosses (by double-click) in "List1_Half_Diallel_from_QCS".
3. Click on the "Option…" button and select the half-diallel method: Number of crosses = Maximum, Contribution of

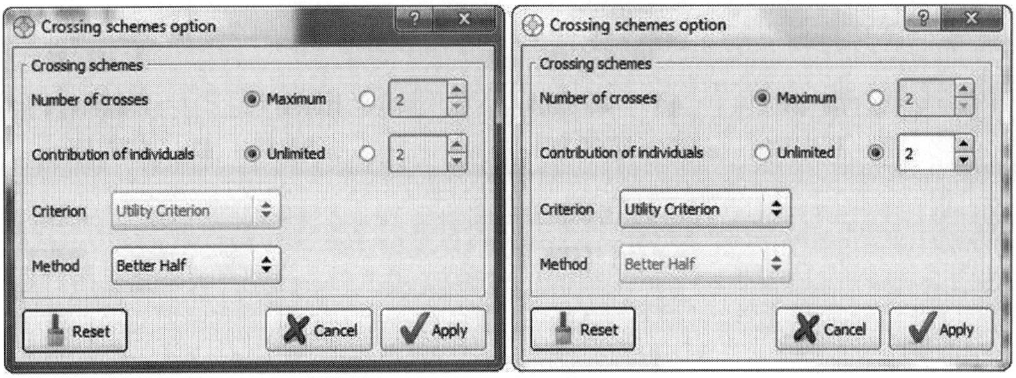

Fig. 9 Crossing schemes option window. *Left*: better-half strategy. *Right*: use of the utility criterion when a constraint is applied on the contribution of parents

individuals = Unlimited, Method = Complete. Notice that the "Criterion" option is disabled by using "maximum"/"unlimited" values for the first two options, Fig. 9. Press "Apply" to validate this strategy.

4. Choose the list of selected candidates coming from the **step 2** (i.e., List(s) selection = List4_Complementation_Selection) and the adequate list of crosses (i.e., List crosses = List1_Half_Diallel_from_QCS). Press the "Run" button to see the results of these crosses stored to the appropriate list.
5. Create a new list named "List2_Better_Half_from_QCS" and do the same procedure with the better-half method.

3.6.2 Constraints: Contribution of Parents/Maximum Number of Crosses

Constraints on the contribution of parents or on the maximum number of crosses to be done can be applied (Figs. 9 and 10). In this case crosses determination is optimized based on the expected (weighted) molecular score or UC of the crosses.

1. Press the "Add" button to create a new list of crosses named "List3_UC_Contrib_2."
2. Press the "Option…" button to open the "Crossing schemes option" window (Fig. 9). Leave Number of crosses = Maximum (by default) and select Contribution of individuals = 2 to specify that each of the selected candidates can be crossed only twice and that there is no constraint on the maximum number of crosses to be made. The criterion box (disabled by default) will appear as in Fig. 9. Select Criterion = Utility Criterion for the first list and press the "Apply" button.
3. Choose the list of selected candidates coming from the **step 2** (i.e., List selection(s) = List4_Complementation_Selection) and

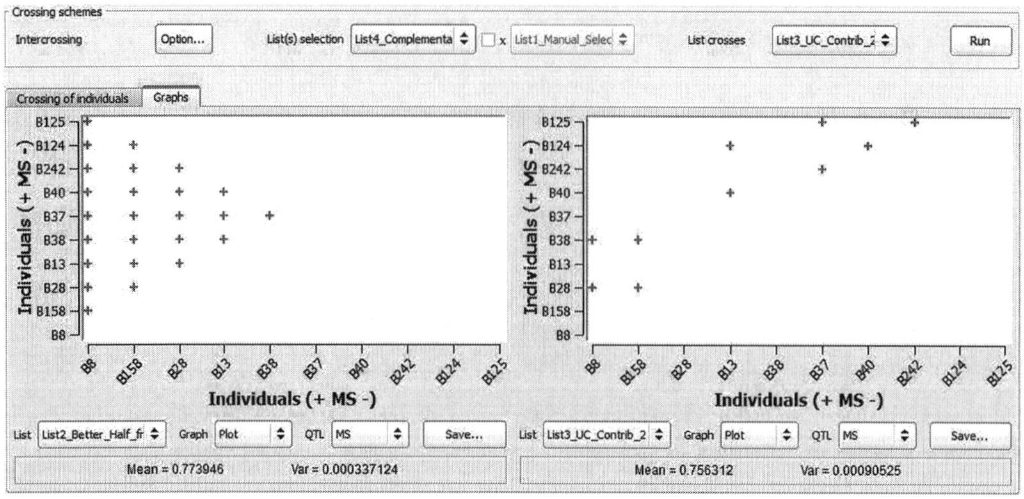

Fig. 10 Display of two mating designs (List2_Better_Half_from_QCS vs. List3_UC_Contrib_2)

select the list where crosses will be stored (i.e., List crosses = List3_UC_Contrib_2. Press the "Run" button to display the results of these crosses.

3.6.3 Factorial Design

Two complementary lists of selected plants can be crossed according to a factorial design.

1. Press the "Add" button to create a new list of crosses and rename it "List4_Factorial_Design."
2. Check the box between the two lists of selection to enable the possibility to select a second list of selected candidates.
3. Select List(s) selection = List1_Manual_Selection x List2_Truncation_MS_Selection and List crosses = List4_Factorial_Design.
4. Press the "Run" button. The list containing only two individuals (B124 and B125) previously found and added via the QCS method has been crossed with the ten individuals having the highest MS resulting in a factorial design (twenty crosses done).

3.6.4 Display and Comparison of Lists of Crosses

As for selected individuals, crosses can be compared through tables and graphs.

1. Click on the "Crossing of individuals" tab to compare two lists of crosses via parallel tables. Left table (i.e., List = List1_Half_diallel_From_QCS) displays the outcome of the "complete" option. OptiMAS computes the expected molecular score of the progeny of each of the 45 crosses, for all and each QTL. Right table (i.e., List = List2_Better_Half_from_QCS) displays the outcome of the "better-half" option, which leads to 25 crosses.

QTL columns can still be sorted, in order for instance to identify pairs having a score of zero. These crosses can be deleted (right click on the selected pair and press the "delete…" button).

2. Click on the "Graphs" tab. Graphs have been automatically generated to display a view of the selected crosses based on the different strategies. Individuals are ranked on the two axes based on their genetic value (MS from highest to lowest). Left side graph displays the outcome of the better-half procedure (List = List2_Better_Half_from_QCS). It illustrates that crosses between individuals having lowest MS have been avoided (i.e., B37 to B125). Right side graph (List = List3_UC_Contrib_2) illustrates the outcome of selection of crosses considering constraints and the utility criterion (Fig. 10).
3. Select the histogram representation (i.e., Graph = Histo) to analyze and compare the lists of crosses for the molecular score at each/all QTL position(s).

4 Notes

1. Two example datasets (moreau/blanc.dat and moreau/blanc.map) are supplied within the user's home directory "/user_name/OptiMAS/input/" for biparental and multiparental designs. Both files must be in plain-text, tab-delimited format (no space between fields) and the markers present in the map file must be ordered and match those in the genotypes/pedigree file.
2. Homozygous genotypes for an allele (e.g., A) can be scored either as A or A/A. Heterozygotes are expected to be separated by a "/" (e.g., A/G). Heterozygous genotypes are assumed to be unphased (i.e., A/B equivalent to B/A). Missing data at marker loci are allowed and must be entered as "-" (or can be left blank). For dominant markers, assuming A dominant vs. *a* recessive, genotypes presenting allele A must be coded A/-. Parental inbred lines should not contain missing data.
3. Never change the header field names (Locus/Chr/QTL/Pos/All+). It is recommended to code parental alleles ("All + " column) by only one character (e.g., a, A, b, 1 …). Check that decimals are 0.00 and not 0,00 for the loci/QTL positions column (Pos). Remember that both the map and the genotypes/pedigree files must be in plain-text, tab-delimited format. So, use tabulation and not spaces between fields. The markers present in the map file must be ordered and match those in the genotypes/pedigree file (same number of markers, only QTL are added). The number of markers selected per QTL should be in the range of 2–7 in order to avoid intensive

computation time (more progress in this area will be made in the next versions).

4. Advanced options/parameters. (a) QTL analyzed: by default all the QTL present in the input files will be analyzed. You can also choose to select a specific QTL to run the analysis. (b) Cut-off—Diplotypes is the genotypic probability below which a rare phased genotype (diplotype) is removed and no more considered in subsequent computations (default value = 0.000001). A diplotype is defined as the union of a pair of unambiguous haplotypes corresponding to parental gametes. (c) Cut-off—Gametes corresponds to the probability that the number of crossovers expected in the region between flanking markers exceeds a given value. Unlikely gametes with number of crossovers over this value are removed and no more considered in subsequent computations. Use of this option with values up to 0.01 is recommended in case many flanking markers per QTL lead to high computation time with default option.

5. Questionable results are displayed in the file "output_folder/date/events_summary.log." It is recommended to always check it before any interpretation. An empty file means that no errors have been found during the execution. Otherwise, this file contains the name of individuals with questionable genotypes at considered QTL positions. These are defined as individual × QTL combinations for which no diplotype passing the cut-off threshold could be found. Three main causes can be distinguished: (a) individuals have genotype data inconsistent in light of their ancestry (which may reveal either a genotyping error or an error in the declared pedigree); (b) individuals display very unlikely genotypes relative to their parents (i.e., genotypes that can only be obtained assuming unlikely recombination events), and (c) the cut off (threshold value) set by user for keeping diplotypes is too high. Note that a genotyping error at generation n affects the results for its progeny at generation $n+1$, even if individuals at this generation have correct genotypes.

6. The "Utility Criterion" is based on the estimation of the expected number of favorable alleles carried by the superior 5 % gametes produced by the individual. For a same MS, this criterion favors individuals with no unfavorable alleles fixed. This score ranges from zero to the number of QTL. The present version of UC estimation assumes independence between QTL and should be considered as only indicative in case of linked QTL. It also assumes that the distribution of scores can be approximated by a normal distribution (which is not valid in case of small number of heterozygous QTL).

7. For example, individual B8 is considered as homozygous favorable for eight QTL (in blue, No.(+/+) = 8), homozygous unfavorable for only one QTL (in red, No.(−/−) = 1), it presents no heterozygous QTL (in grey, No.(+/−) = 0) and two QTL (in yellow, No.(?) = 2) are uncertain. Some MS at QTL positions are close to 1 (e.g., QTL3 = 0.9761) and some others are lower (e.g., QTL2 = 0.8598). This uncertainty can be due to (a) the fact that one marker flanking the QTL is heterozygous, whereas the other one is homozygous favorable, which indicates that a recombination took place near the QTL position, or (b) that there are missing data (*see* genotypes/pedigree file).
8. The QCS approach can be applied to any predefined list (including the possibility to consider an empty list). In this example, the first eight individuals ($N_0 = 8$) were selected via the truncation selection (based on the MS criterion). Then (a) the QTL for which the favorable alleles are "present" in fewer than n_t selected individuals are identified (Fig. 4, in red); (b) among the remaining individuals (belonging to Cycle 2, C2), taken in order of decreasing MS (with $MS \geq MS_{min}$), OptiMAS searches for the individual having favorable alleles "present" at the largest number of those QTL identified in (a). This individual is added to the subset of selected individuals (i.e., $N = N_0 + 1$). Steps (a) and (b) are iterated until either of the following conditions is met: favorable alleles at all QTL are present in at least n_t individuals of the selected subset, or the number of individuals in the selected subset reaches the given N_{max} value, or it is not possible to find an individual in step (b). Notice that in step (b), MS is a secondary criterion; individuals are taken based on their ability to complement the subset of already selected individuals.
9. The pedigree representation is useful to follow the contribution of selected individuals over generations of selection and to prevent possible bottlenecks (individuals coming from a reduced number of parents at a given generation), in order to limit risk of drift which may lead for instance to the fixation of an undesired phenotypic type for traits not considered in the MARS process. It also can be used to maintain diversity for selection on traits complementary to those considered for the MARS process.

Acknowledgements

We are grateful to colleagues within INRA and the Generation Challenge Programme (GCP) for enlightening user oriented input to this project. This development has benefited from the advices and beta-testing of Delphine Fleury and Mark Sawkins. This program is part of the IBP within GCP and is funded by the Bill and Melinda Gates Foundation.

References

1. Varshney RK, Hoisington DA, Tyagi AK (2006) Advances in cereal genomics and applications in crop breeding. Trends Biotechnol 24: 490–499
2. Lande R, Thompson R (1990) Efficiency of marker-assisted selection in the improvement of quantitative traits. Genetics 124:743–756
3. Xu Y, Crouch JH (2008) Marker-assisted selection in plant breeding: from publications to practice. Crop Sci 48:391–407
4. Rebai A, Goffinet B (2000) More about quantitative trait locus mapping with diallel design. Genet Res 75:243–247
5. Blanc G, Charcosset A, Mangin B, Gallais A, Moreau L (2006) Connected populations for detecting quantitative trait loci and testing for epistasis: an application in maize. Theor Appl Genet 113:206–224
6. Blanc G, Charcosset A, Veyrieras JB, Gallais A, Moreau L (2008) Marker-assisted selection efficiency in multiple connected populations: a simulation study based on the results of a QTL detection experiment in maize. Euphytica 161: 71–84
7. Yu J, Holland JB, McMullen MD, Buckler ES (2008) Genetic design and statistical power of nested association mapping in maize. Genetics 178:539–551
8. Valente F, Gauthier F, Bardol N, Blanc G, Joets J, Charcosset A, Moreau L (2013) OptiMAS: a decision support tool for marker-assisted assembly of diverse alleles. J Hered 104:586–590
9. Hospital F, Goldringer I, Openshaw S (2000) Efficient marker-based recurrent selection for multiple quantitative trait loci. Genet Res 75:357–368
10. Bernardo R, Moreau L, Charcosset A (2006) Number and fitness of selected individuals in marker-assisted and phenotypic recurrent selection. Crop Sci 46:1972–1980

Chapter 10

Genomic Selection in Plant Breeding

Mark A. Newell and Jean-Luc Jannink

Abstract

Genomic selection (GS) is a method to predict the genetic value of selection candidates based on the genomic estimated breeding value (GEBV) predicted from high-density markers positioned throughout the genome. Unlike marker-assisted selection, the GEBV is based on all markers including both minor and major marker effects. Thus, the GEBV may capture more of the genetic variation for the particular trait under selection.

Key words Genomic selection, Genomic estimated breeding value, Population structure, Training population, Selection candidates, Cross-validation

1 Introduction

Phenotypic selection is challenging for three reasons. First, many phenotypes targeted by breeders are slow to measure, either because they require plots with many plants to be measured and therefore time for extensive propagation of the genetic individual, or they require advanced-stage inbred lines and therefore time for inbreeding, or they express themselves late in the life cycle of the plant. Second, they are expensive to measure. Third, compounding the expense, experimental error and environment effects often make the phenotype an imperfect guide to the potential of the underlying genotype and therefore the phenotypic measurement needs to be replicated. All of these challenges would fall away if we could infer the phenotype from the genotype: the genotype can be assessed before a seed is even germinated, the cost of genotyping continues to plummet, and measurement error on the genotype is usually quite small. Meuwissen et al. [1] first proposed a series of methods to predict future phenotype on the basis of genotype and dubbed these methods "genomic selection."

Delphine Fleury and Ryan Whitford (eds.), *Crop Breeding: Methods and Protocols*, Methods in Molecular Biology, vol. 1145, DOI 10.1007/978-1-4939-0446-4_10, © Springer Science+Business Media New York 2014

In genomic selection, a training population that contains individuals that have both phenotypic and genotypic data is used to fit a prediction model. That model is then used to predict the breeding value or the genotypic value of individuals that have only genotypic data. Selection can then proceed on the basis of these predictions. The first evidence that genomic selection could work empirically was obtained in dairy cattle [2]. Similar evidence is now available for crops [3–6]. Since then, research has clarified the relationships between trait heritability, marker density, training population size, and the effective population size of the breeding program in determining the accuracy of prediction [7, 8]. We also know more about how the genetic relatedness between individuals in the training and prediction populations affects accuracy [9–11], and of the impact of population structure on the prediction process [6, 12, 13, 14]. Finally, simulations have begun to look at how genomic selection might affect long-term gains [15–17]. Thus, the way forward to develop optimal breeding schemes from the perspective of short-term and long-term gain as well as economic efficiency [18] is becoming clearer. In addition, software to facilitate prediction is becoming available [19].

The objective of this chapter is to walk the reader through the analysis of genomic prediction on a very simple dataset. For a complete beginner in R, we show how to install R and the extra R packages that are needed for the analyses. We show how to load the data into R, to have a quick look at it, and to run a basic analysis. For users more familiar with R, we provide code to run and interpret a cross-validation of the prediction model, to explore the impact of the training population size on prediction accuracy, and to have a look at subpopulation structure within the training population. In the beginning of the chapter, all statements in the R script "GSChapterCode.R" are shown with the output that they produce in screen shots. Later in the chapter as we assume users become more familiar with R statements, only select statements and their output are given in the text. The accompanying script, however, is well documented. All files, including "GSChapterCode.R," can be downloaded from https://www.integratedbreeding.net/ib-tools/genomicselection.

2 Software Installation

2.1 R Installation

The R software is a free and open-source statistical software that can be downloaded for Linux, MacOS X, or Windows from www.r-project.org. From the Web site, click on CRAN (Comprehensive R Archive Network), select the appropriate CRAN Mirror, and select the appropriate operating system. Follow on-screen prompts for installation.

3 Install and Load Packages

1. *Install Packages*

Two packages will be used in this chapter, rrBLUP [19] for implementation of GS and ggplot2 [20] for data visualization. R packages can be either installed from the "Packages" drop-down menu within R or by using the install.packages() function. The latter will be used here for demonstration, within the R console window:

```
> install.packages("rrBLUP")
```

After the first install.packages() line is run the "CRAN mirror" window will appear where the appropriate CRAN mirror should be selected (Fig. 1).

After choosing the appropriate CRAN mirror, installation should take place and the following should appear in the R Console indicating the location of the downloaded packages (Fig. 2).

Follow the same instructions for installation of the ggplot2 package:

```
> install.packages("ggplot2")
```

2. *Load Installed Packages*

Once the two packages are installed they need to be loaded into the R software. This is done by using the library() function within the R Console as follows:

```
> library(rrBLUP)
> library(ggplot2)
```

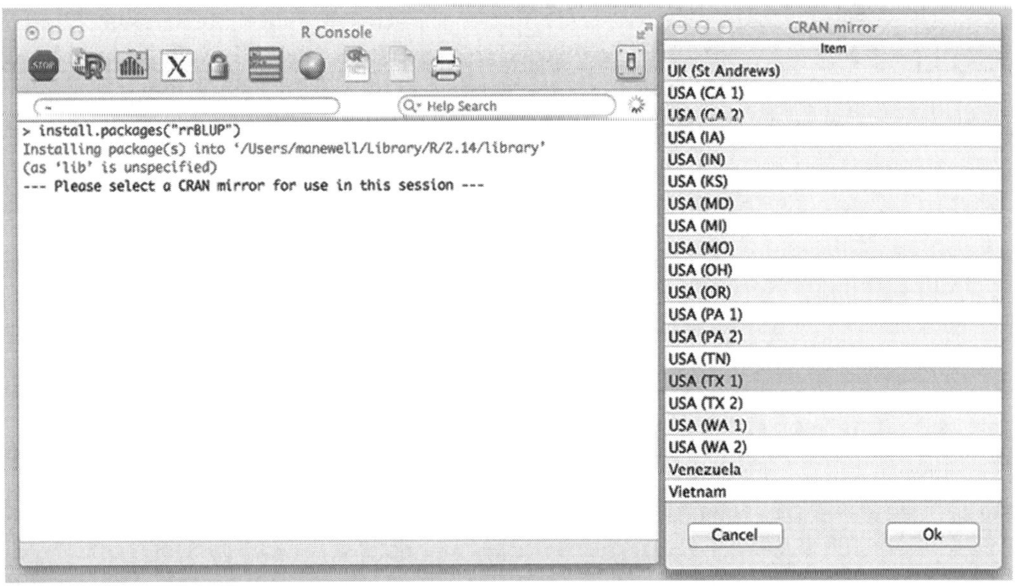

Fig. 1 Screenshot of the install.packages() line to select the "CRAN mirror"

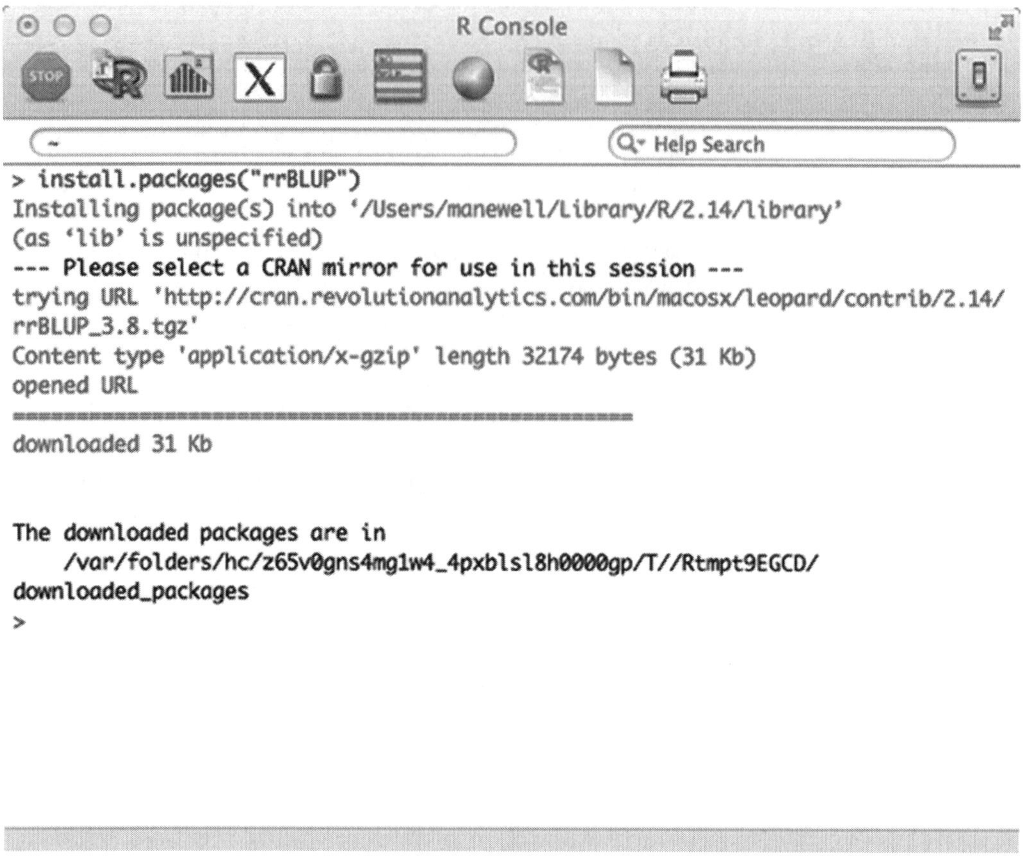

Fig. 2 Screenshot of the R console indicating the location of the downloaded packages

When the R software is closed and reopened the library() function needs to be run for all packages that will be used unless they are within the base R functions. The base R functions include simple functions such as mean() for calculating the mean or plot() for basic plotting. However, the install.packages() function only needs to be run during the initial installation of packages into R.

4 Data for Genomic Selection

Genomic selection consists of a training set with both phenotypic and genotypic data; this set is used to predict marker effects. The second group, referred to as the selection candidates, includes lines that have been genotyped but have unknown phenotype. The overall goal of GS is to predict the genomic estimated breeding value (GEBV) for the selection candidates based on the predicted marker effects of the training set. Here we will use a barley (*Hordeum vulgare* L.) dataset with phenotypic data for Fusarium Head Blight (FHB) resistance [6]. The dataset consists of 3 years

(2006, 2007, and 2008) and three institutions (Busch Agricultural Resources (BA), The University of North Dakota (N6), and The University of Minnesota (MN)).

Breeding programs typically operate such that the earlier material is used as the training population to predict the GEBV of later material, or selection candidates. Following the same scheme here, the first two years (2006 and 2007) will be used as the training population and the last year (2008), the selection candidates. For simplicity, only the lines from The University of Minnesota will be used in the following example.

4.1 Choose Working Directory

Four files are required for the following example including *FHBtraining.csv*, *FHBcandidates.csv*, *GENOtraining.csv*, and *GENOcandidates.csv* and should be downloaded (from https://www.integratedbreeding.net/ib-tools/genomicselection) and placed in a single folder on the working computer. The next step is to choose the working directory within R in which files are called. From the File menu select Change dir and set the appropriate file in which the four files have been saved. This enables reading and writing files to the particular file on the working computer.

4.2 Upload Training Population Data

The training population data includes both the phenotypic data for FHB entitled *FHBtraining.csv* and the genotypic data *GENOtraining.csv*. Both of the files can be uploaded in a similar fashion using the read.csv() function within base R. The <- designates the name of the dataset within R.

```
> FHBtraining <-read.csv(file="FHBtraining.csv",row.names=1)
> GENOtraining<-read.csv(file="GENOtraining.csv",row.names=1)
```

The read.csv() function reads only from the working directory and row.names=1 indicates that the first column is composed of row names for the data.

4.3 Upload Selection Candidates Data

The same functions can be used to upload the data for the selection candidates. In most cases only the genotypic data for the selection candidates would be used; however, for this example the phenotypic data for the selection candidates will also be uploaded for evaluation purposes.

```
>FHBcandidates<-read.csv(file="FHBcandidates.csv",row.names=1)
>GENOcandidates<-read.csv(file="GENOcandidates.csv",row.names=1)
```

4.4 Exploring the Data

The data can then be explored with various functions within base R, typing only the name of the dataset prints that particular dataset:

```
> head(FHBtraining)
> GENOtraining[1:6,1:6]
```

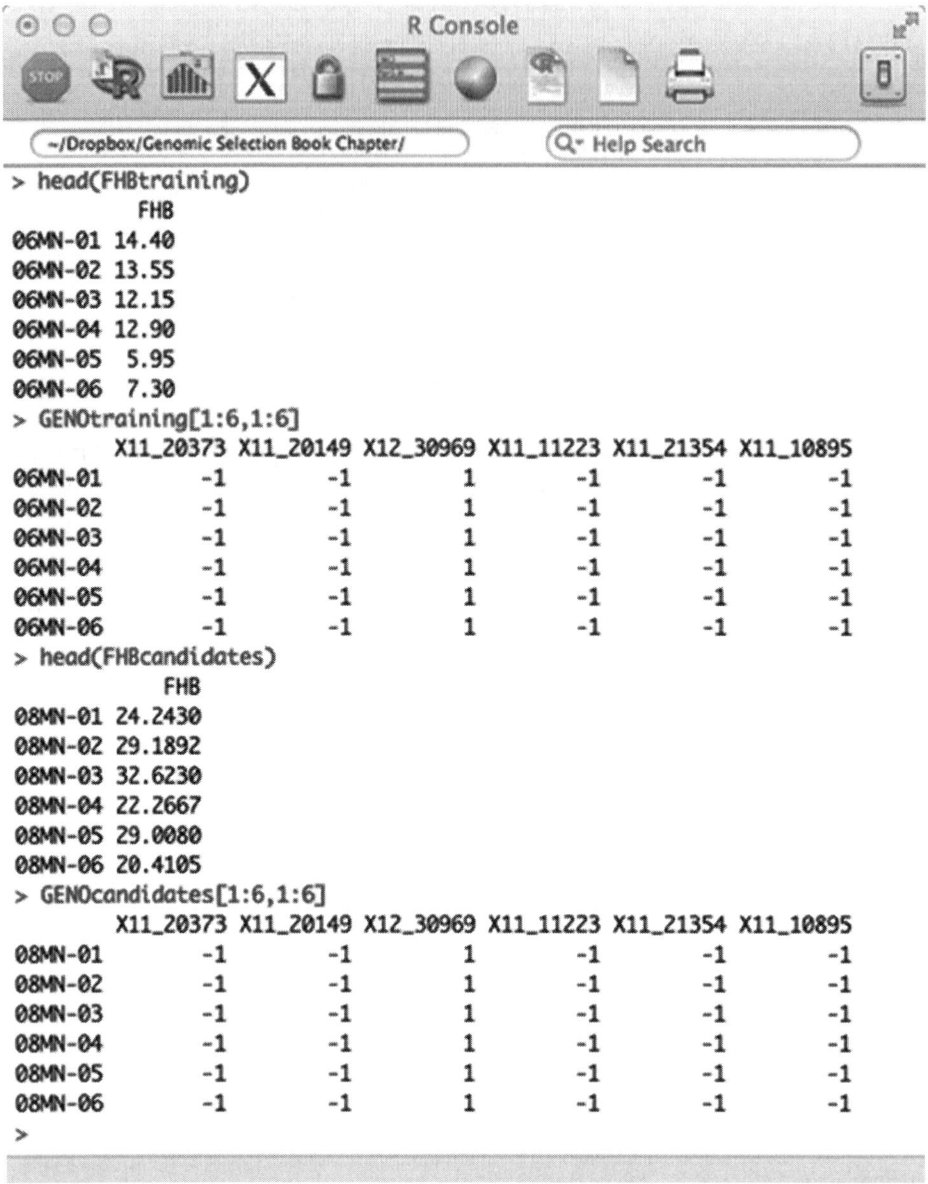

Fig. 3 Screenshot of the R console showing the phenotyping and genotyping data

```
> head(FHBcandidates)
> GENOcandidates[1:6,1:6]
```

For the phenotypic data, the head() function has been used to only print the first six columns. Likewise, brackets have been used to print the first six rows and columns of the genotypic datasets. The output from the above code is shown in Fig. 3. It can be seen that the phenotypic datasets are vectors and the genotypic datasets are data matrices where the rows represent the lines and columns represent single nucleotide polymorphism (SNP) markers in the format −1, 0, and 1 for aa, Aa, and AA, respectively.

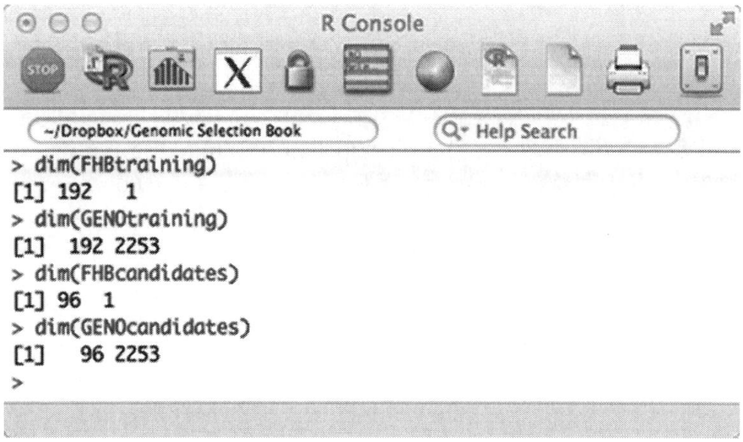

Fig. 4 Screenshot of the R console showing the row by column dimension of the phenotyping and genotyping datasets

Another useful function that supplies important information concerning the datasets is the dim() function which prints the row by column dimensions.

```
> dim(FHBtraining)
> dim(GENOtraining)
> dim(FHBcandidates)
> dim(GENOcandidates)
```

It can be seen from the output that there are 192 lines in the training population and 96 lines in the selection candidates (Fig. 4). The genotypic datasets contain 2253 SNP markers.

4.5 Population Structure

Population structure or differing levels of genetic relatedness among the individuals in a population has been demonstrated to be major concern for genome-wide methods and can lead to spurious associations in genome-wide association studies (GWAS, [21]). Although this is a major concern for GWAS, its relevance for GS shifts more towards maintaining accuracy across subpopulation structure [21].

There are many methods to deal with subpopulation structure for GS. We discuss the effect of population structure on accuracy at the end of the chapter but for simplicity we will ignore it in our initial demonstration of GS here.

5 Genomic Selection

The rrBLUP package was installed and loaded into R in Subheading 3.1 and will be used to predict the GEBVs for the selection candidates. This package has many functions that can be further explored on the R Web site at http://cran.r-project.org/web/packages/rrBLUP/index.html. The reference manual can also be found on the Web site which includes a summary of the package, the name of the developer, any relevant publications, and all of its functions.

5.1 Predict GEBV for Selection Candidates

The kinship.BLUP() function within the rrBLUP package implements ridge regression–best linear unbiased prediction for GEBV. In order to see the R help window for any function a question mark (?) followed by the package name can be entered into the console:

```
> ?kinship.BLUP
```

The help window shows the function with arguments that need to be passed in order to use the kinship.BLUP() function (Fig. 5). The "Value" section within the help window shows the information included in the output file including the BLUP solutions for both the training ($g.train) and selection candidates ($g.pred). The following code is used to implement GS for the FHB barley data:

```
>GEBV<-kinship.BLUP(y=as.matrix(FHBtraining),
G.train=as.matrix (GENOtraining),G.pred=as.matrix
(GENOcandidates))
```

where y is the phenotypic data for the training population, G.train is the genotypic data for the training population, and G.pred is the genotypic data for the selection candidates. The as.matrix() function is used for all of the inputs, this is required by the kinship. BLUP() function.

5.2 The GEBV Output

The output from the kinship.BLUP() function was stored in a dataset called GEBV, the contents of GEBV can be printed by typing GEBV in the R console:

```
> GEBV
```

There are three components of the output, the BLUP solution for the training set ($g.train), the RR–BLUP solution for the selection candidates ($g.pred), and the estimate for the fixed effects ($beta), which in this case is the intercept (Fig. 6).

5.3 Accuracy of GS

1. *Calculate and validate the accuracy of GS*
   The accuracy of GS can be calculated as the correlation between the true and GEBV for the selection candidates. We can also calculate the accuracy of the GEBV for the training set. Using the cor() function within base R these can easily be calculated as follows:

   ```
   > cor(GEBV$g.pred,FHBcandidates)
   > cor(GEVB$g.train,FHBtraining)
   ```

   In this case, GEBV$g.pred calls the GEBV for the selection candidates and FHBcandidates are the true breeding values for the selection candidates. In this situation, the correlation between the prediction and the candidate phenotypes is 0.58 (Fig. 7).

R Help

kinship.BLUP {rrBLUP} R Documentation

Genomic prediction by kinship-BLUP

Description

This function uses mixed.solve for genomic prediction based on the kinship between lines.

Usage

```
kinship.BLUP(y, G.train, G.pred=NULL, X=NULL, Z.train=NULL,
     K.method="RR", n.profile=10, mixed.method="REML", n.core=1)
```

Arguments

y	Vector (*n.obs* \times *1*) of observations. Missing values (NA) are omitted (see mixed.solve).
G.train	Matrix (*n.train* \times *m*) of unphased genotypes for the training population: *n.train* lines with *m* bi-allelic markers. Genotypes should be coded as {-1,0,1} = {aa,Aa,AA}; fractional (imputed) and missing (NA) alleles are allowed.
G.pred	Matrix (*n.pred* \times *m*) of unphased genotypes for the prediction population: *n.pred* lines with *m* bi-allelic markers. Genotypes should be coded as {-1,0,1} = {aa,Aa,AA}; fractional (imputed) and missing (NA) alleles are allowed.
X	Design matrix (*n.obs* \times *p*) of fixed effects. If not passed, a vector of 1's is used to model the intercept.
Z.train	0-1 matrix (*n.obs* \times *n.train*) relating observations to lines in the training set. If not passed the identity matrix is used.
K.method	"RR" (default) is ridge regression, for which K is the realized additive relationship matrix computed with A.mat. The option "GAUSS" is a Gaussian kernel ($K = e^{-D^2/\theta^2}$) and "EXP" is an exponential kernel ($K = e^{-D/\theta}$), where Euclidean distances *D* are computed with dist.
n.profile	For K.method = "GAUSS" or "EXP", the number of points to use in the log-likelihood profile for the scale parameter θ.
mixed.method	Either "REML" (default) or "ML".
n.core	For Mac, Linux, and UNIX users, setting n.core > 1 will enable parallel execution on a machine with multiple cores. R package multicore must be installed for this to work. Do not run multicore from within the R GUI; you must use the command line.

Value

$g.train

BLUP solution for the training set

$g.pred

BLUP solution for the prediction set (when G.pred != NULL)

$beta

ML estimate of fixed effects

For GAUSS or EXP, function also returns

$profile

log-likelihood profile for the scale parameter

References

Endelman, J.B. 2011. Ridge regression and other kernels for genomic selection with R package rrBLUP. Plant Genome 4:250-255.

Fig. 5 Help window of the kinship.BLUP() function

A visual representation of the accuracy can also be printed using the ggplot2 package as follows:

```
> qplot(x=FHBcandidates$FHB,y=GEBV$g.pred)+
geom_smooth(method="lm")
```

The GEBV for the training set is 0.85 (Fig. 8). Note how much higher this correlation is than was the correlation on the candidate population. For the training population, the prediction model is "cheating": the data that were used to fit the prediction

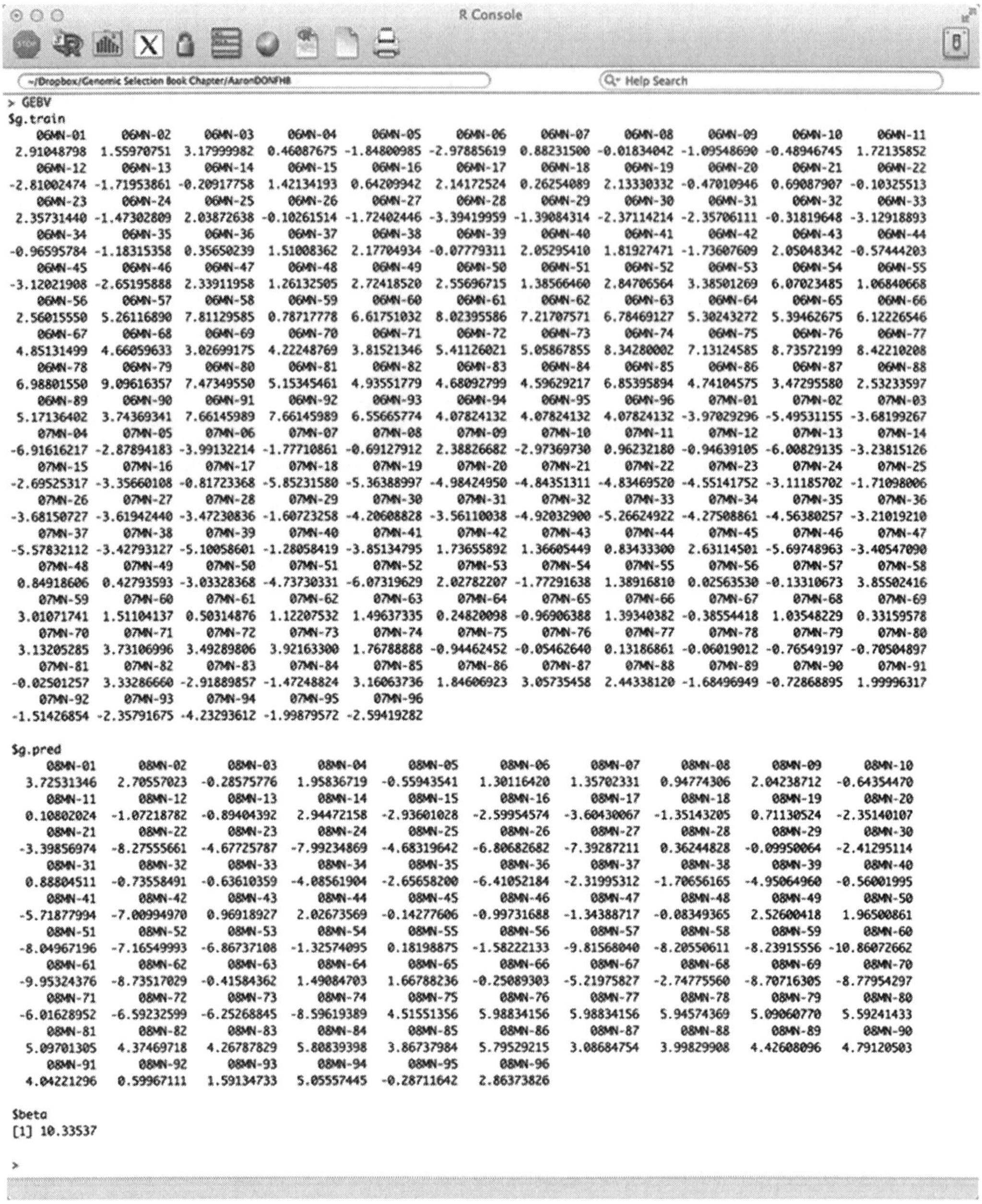

```
> GEBV
$g.train
    06MN-01     06MN-02     06MN-03     06MN-04     06MN-05     06MN-06     06MN-07     06MN-08     06MN-09     06MN-10     06MN-11
 2.91048798  1.55970751  3.17999982  0.46087675 -1.84800985 -2.97885619  0.88231500 -0.01834042 -1.09548690 -0.48946745  1.72135852
    06MN-12     06MN-13     06MN-14     06MN-15     06MN-16     06MN-17     06MN-18     06MN-19     06MN-20     06MN-21     06MN-22
-2.81002474 -1.71953861 -0.20917758  1.42134193  0.64209942  2.14172524  0.26254089  2.13330332 -0.47010946  0.69087907 -0.10325513
    06MN-23     06MN-24     06MN-25     06MN-26     06MN-27     06MN-28     06MN-29     06MN-30     06MN-31     06MN-32     06MN-33
 2.35731440 -1.47302809  2.03872638 -0.10261514 -1.72402446 -3.39419959 -1.39084314 -2.37114214 -2.35706111 -0.31819648 -3.12918893
    06MN-34     06MN-35     06MN-36     06MN-37     06MN-38     06MN-39     06MN-40     06MN-41     06MN-42     06MN-43     06MN-44
-0.96595784 -1.18315358  0.35650239  1.51008362  2.17704934 -0.07779311  2.05295410  1.81927471 -1.73607609  2.05048342 -0.57444203
    06MN-45     06MN-46     06MN-47     06MN-48     06MN-49     06MN-50     06MN-51     06MN-52     06MN-53     06MN-54     06MN-55
-3.12021908 -2.65195888  2.33911958  1.26132505  2.72418520  2.55696715  1.38566460  2.84706564  3.38501269  6.07023485  1.06840668
    06MN-56     06MN-57     06MN-58     06MN-59     06MN-60     06MN-61     06MN-62     06MN-63     06MN-64     06MN-65     06MN-66
 2.56015550  5.26116890  7.81129585  0.78717778  6.61751032  8.02395586  7.21707571  6.78469127  5.30243272  5.39462675  6.12226546
    06MN-67     06MN-68     06MN-69     06MN-70     06MN-71     06MN-72     06MN-73     06MN-74     06MN-75     06MN-76     06MN-77
 4.85131499  4.66059633  3.02699175  4.22248769  3.81521346  5.41126021  5.05867855  8.34280002  7.13124585  8.73572199  8.42210208
    06MN-78     06MN-79     06MN-80     06MN-81     06MN-82     06MN-83     06MN-84     06MN-85     06MN-86     06MN-87     06MN-88
 6.98801550  9.09616357  7.47349550  5.15345461  4.93551779  4.68092799  4.59629217  6.85395894  4.74104575  3.47295580  2.53233597
    06MN-89     06MN-90     06MN-91     06MN-92     06MN-93     06MN-94     06MN-95     06MN-96     07MN-01     07MN-02     07MN-03
 5.17136402  3.74369341  7.66145989  7.66145989  6.55665774  4.07824132  4.07824132  4.07824132 -3.97029296 -5.49531155 -3.68199267
    07MN-04     07MN-05     07MN-06     07MN-07     07MN-08     07MN-09     07MN-10     07MN-11     07MN-12     07MN-13     07MN-14
-6.91616217 -2.87894183 -3.99132214 -1.77710861 -0.69127912  2.38826682 -2.97369730  0.96232180 -0.94639105 -6.00829135 -3.23815126
    07MN-15     07MN-16     07MN-17     07MN-18     07MN-19     07MN-20     07MN-21     07MN-22     07MN-23     07MN-24     07MN-25
-2.69525317 -3.35660108 -0.81723368 -5.85231580 -5.36388997 -4.98424950 -4.84351311 -4.83469520 -4.55141752 -3.11185702 -1.71098006
    07MN-26     07MN-27     07MN-28     07MN-29     07MN-30     07MN-31     07MN-32     07MN-33     07MN-34     07MN-35     07MN-36
-3.68150727 -3.61942440 -3.47230836 -1.60723258 -4.20608828 -3.56110038 -4.92032900 -5.26624922 -4.27508861 -4.56380257 -3.21019210
    07MN-37     07MN-38     07MN-39     07MN-40     07MN-41     07MN-42     07MN-43     07MN-44     07MN-45     07MN-46     07MN-47
-5.57832112 -3.42793127 -5.10058601 -1.28058419 -3.85134795  1.73655892  1.36605449  0.83433300  2.63114501 -5.69748963 -3.40547090
    07MN-48     07MN-49     07MN-50     07MN-51     07MN-52     07MN-53     07MN-54     07MN-55     07MN-56     07MN-57     07MN-58
 0.84918606  0.42793593 -3.03328368 -4.73730331 -6.07319629  2.02782207 -1.77291638  1.38916810  0.02563530 -0.13310673  3.85502416
    07MN-59     07MN-60     07MN-61     07MN-62     07MN-63     07MN-64     07MN-65     07MN-66     07MN-67     07MN-68     07MN-69
 3.01071741  1.51104137  0.50314876  1.12207532  1.49637335  0.24820098 -0.96906388  1.39340382 -0.38554418  1.03548229  0.33159578
    07MN-70     07MN-71     07MN-72     07MN-73     07MN-74     07MN-75     07MN-76     07MN-77     07MN-78     07MN-79     07MN-80
 3.13205285  3.73106996  3.49289806  3.92163300  1.76788888 -0.94462452 -0.05462640  0.13186861 -0.06019012 -0.76549197 -0.70504897
    07MN-81     07MN-82     07MN-83     07MN-84     07MN-85     07MN-86     07MN-87     07MN-88     07MN-89     07MN-90     07MN-91
-0.02501257  3.33286660 -2.91889857 -1.47248824  3.16063736  1.84606923  3.05735458  2.44338120 -1.68496949 -0.72868895  1.99996317
    07MN-92     07MN-93     07MN-94     07MN-95     07MN-96
-1.51426854 -2.35791675 -4.23293612 -1.99879572 -2.59419282

$g.pred
     08MN-01      08MN-02      08MN-03      08MN-04      08MN-05      08MN-06      08MN-07      08MN-08      08MN-09      08MN-10
  3.72531346   2.70557023  -0.28575776   1.95836719  -0.55943541   1.30116420   1.35702331   0.94774306   2.04238712  -0.64354470
     08MN-11      08MN-12      08MN-13      08MN-14      08MN-15      08MN-16      08MN-17      08MN-18      08MN-19      08MN-20
  0.10802024  -1.07218782  -0.89404392   2.94472158  -2.93601028  -2.59954574  -3.60430067  -1.35143205   0.71130524  -2.35140107
     08MN-21      08MN-22      08MN-23      08MN-24      08MN-25      08MN-26      08MN-27      08MN-28      08MN-29      08MN-30
 -3.39856974  -8.27555661  -4.67725787  -7.99234869  -4.68319642  -6.80682682  -7.39287211   0.36244828  -0.09950064  -2.41295114
     08MN-31      08MN-32      08MN-33      08MN-34      08MN-35      08MN-36      08MN-37      08MN-38      08MN-39      08MN-40
  0.88804511  -0.73558491  -0.63610359  -4.08561904  -2.65658200  -6.41052184  -2.31995312  -1.70656165  -4.95064960  -0.56001995
     08MN-41      08MN-42      08MN-43      08MN-44      08MN-45      08MN-46      08MN-47      08MN-48      08MN-49      08MN-50
 -5.71877994  -7.00994970   0.96918927   2.02673569  -0.14277606  -0.99731688  -1.34388717  -0.08349365   2.52600418   1.96500861
     08MN-51      08MN-52      08MN-53      08MN-54      08MN-55      08MN-56      08MN-57      08MN-58      08MN-59      08MN-60
 -8.04967196  -7.16549993  -6.86737108  -1.32574095   0.18198875  -1.58222133  -9.81568040  -8.20550611  -8.23915556 -10.86072662
     08MN-61      08MN-62      08MN-63      08MN-64      08MN-65      08MN-66      08MN-67      08MN-68      08MN-69      08MN-70
 -9.95324376  -8.73517029  -0.41584362   1.49084703   1.66788236  -0.25089303  -5.21975827  -2.74775560  -8.70716305  -8.77954297
     08MN-71      08MN-72      08MN-73      08MN-74      08MN-75      08MN-76      08MN-77      08MN-78      08MN-79      08MN-80
 -6.01628952  -6.59232599  -6.25268845  -8.59619389   4.51551356   5.98834156   5.98834156   5.94574369   5.09060770   5.59241433
     08MN-81      08MN-82      08MN-83      08MN-84      08MN-85      08MN-86      08MN-87      08MN-88      08MN-89      08MN-90
  5.09701305   4.37469718   4.26787829   5.80839398   3.86737984   5.79529215   3.08684754   3.99829908   4.42608096   4.79120503
     08MN-91      08MN-92      08MN-93      08MN-94      08MN-95      08MN-96
  4.04221296   0.59967111   1.59134733   5.05557445  -0.28711642   2.86373826

$beta
[1] 10.33537

>
```

Fig. 6 Output from the kinship.BLUP() function as stored in a dataset called GEBV

model are also being predicted by it. To avoid cheating, use cross-validation: split the training population into n parts. Each part is called a "fold." For each fold in turn remove the fold from the training population, fit the model, and then predict that fold. This process thus avoids cheating. In the R script, we provide code for running cross-validation using a specified number of folds (fivefold or tenfold cross-validation is typical) and repeating the cross-validation a specified number of times. The correlation

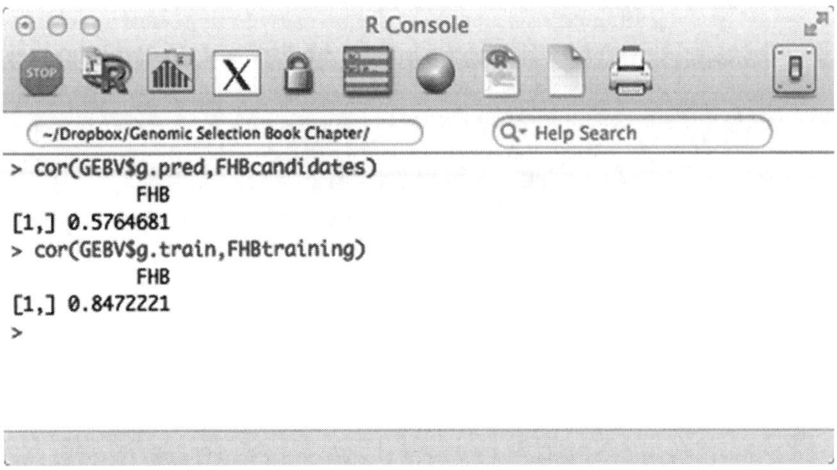

Fig. 7 Correlation values between the prediction and the candidate phenotypes, using GEBV$g.pred for the selection candidates and FHBcandidates for the true breeding values for the selection candidates

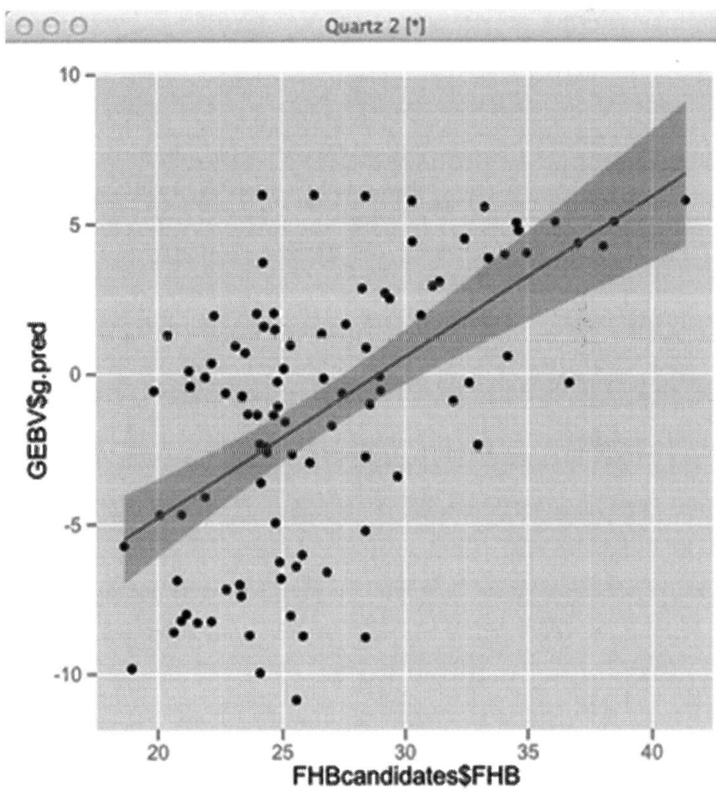

Fig. 8 Visual representation of the accuracy og GS using the ggplot2 package

obtained using cross-validation (given by meanCorrelAcrossRepeats from the function runCrossValidation) should be much closer to that which will be obtained on a true candidate population than if cross-validation is not used.

A further useful parameter provided by cross-validation is the variability of the correlation across the cross-validation folds (given by meanStdDevCorrelAcrossFolds). There is error in the estimate of the correlation that comes from sampling a training population and sampling a validation or candidate population. The standard deviation of the correlation gives the user a measure of how much variation in the correlation is due to sampling. The cross-validation function is:

```
>crossValOut<- runCrossValidation(pheno=F
HBtraining$FHB, geno=GENOtraining, nFolds=5,
nRepeats=10)
> print(crossValOut$meanCorrelAcrossRepeats)
[1] 0.6495281
> print(crossValOut$meanStdDevCorrelAcrossFo
lds)
[1] 0.09555406
```

The cross-validated correlation is about 0.65 for this dataset, which is indeed closer to the correlation on the candidate individuals (0.58). The standard deviation of the correlation shows that a dataset sampled differently could easily have had a correlation anywhere between 0.55 and 0.75: there is variability in GS correlation estimates.

2. *Effect of Training Population Size on Accuracy*

The training population size can have a large effect on the accuracy of GS. The effect of training population size on accuracy can be tested by drawing random samples of different sizes from the training population and calculating the accuracy. The following function tests the effect of training population size by taking random samples of 8, 16, 32, 64, 128, and the complete dataset of 192 genotypes. The function requires a single argument, Ntimes, which designates the number of times a random sample is taken at each training population size.

```
> TrainingSizeOut<-testTrainingSize(Ntimes=
100)
```

The results can then be visualized by generating a boxplot within the ggplot2 package, demonstrating the expected increase in accuracy as the training population size is increased. Nevertheless, accuracy is quite high even for a training population size of eight. This may be caused by different levels of the trait among subpopulations (see below).

3. *Effect of Population Structure on Accuracy*
   As noted, the effect of population structure for GS shifts more towards maintaining accuracy across subpopulation structure [21]. Sample code is provided that demonstrates the effect of population structure for the Minnesota dataset and is accounted for using principal components analysis (PCA).

```
> mrkPCA<- prcomp(rbind(GENOtraining, GENO
candidates))
> scoresTraining<- mrkPCA$x[1:nrow(GENO
training), 1:2]
> scoresCandidates<- mrkPCA$x[nrow(GENOtrain
ing)+1:nrow(GENOcandidates), 1:2]
```

A regression analysis shows that the second principal component has a large effect on FHB resistance.

```
> pcOnTraining<- lm(FHBtraining$FHB~scoresTr
aining[,1]+scoresTraining[,2])
> anova(pcOnTraining)
Analysis of Variance Table
Response: FHBtraining$FHB
                      Df Sum Sq Mean Sq F value
Pr(>F)
scoresTraining[, 1]   1   30.5   30.51  1.3457
0.2475
scoresTraining[, 2]   1 1499.4 1499.38 66.1429
5.428e-14 ***
Residuals           189 4284.4   22.67
---
Signif. codes:  0 '***' 0.001 '**' 0.01 '*'
0.05 '.' 0.1 ' ' 1
```

Thus, this PC is used as a fixed effect (which is specified by a matrix X) in the kinship.BLUP() function.

```
> GEBVst<- kinship.BLUP(y=FHBtraining$FHB,
G.train=GENOtraining, X=cbind(1, scoresTrain-
ing[,2]), G.pred=GENOcandidates)
```

In this specific example the effect of subpopulation structure on the predictions is large, and thus, this factor can play a major role in the accuracy of GS. Including population structure in the training model and seeking to predict FHB severity outside of that related to structure decreases the accuracy from 0.58 to 0. Essentially all of the variation that GS could explain was accounted for by structure. There are not clear subpopulations, but there is relatively simple genomic differentiation between individuals that are resistant versus susceptible to FHB. This explains why it was possible to capture most of this differentiation with a TP size of only 8 (see above).

Acknowledgments

Funding for the data used in this chapter came from a US Department of Agriculture, National Institute of Food and Agriculture agreement, No. 2006-55606-16722 "Barley Coordinated Agricultural Project: Leveraging Genomics, Genetics, and Breeding for Gene Discovery and Barley Improvement." Mark Newell's contributed work was funded by "The Samuel Roberts Noble Foundation."

References

1. Meuwissen THE, Hayes BJ, Goddard ME (2001) Prediction of total genetic value using genome-wide dense marker maps. Genetics 157:1819–1829
2. VanRaden PM, Van Tassell CP, Wiggans GR, Sonstegard TS, Schnabel RD, Taylor JF, Schenkel FS (2009) Invited review: reliability of genomic predictions for North American Holstein bulls. J Dairy Sci 92:16–24
3. Lorenzana RE, Bernardo R (2009) Accuracy of genotypic value predictions for marker-based selection in biparental plant populations. Theor Appl Genet 120:151–161
4. de los Campos G, Naya G, Gianola D, Crossa J, Legarra A, Manfredi E, Weigel K, Cotes JM (2009) Predicting quantitative traits with regression models for dense molecular markers and pedigrees. Genetics 182:375–385
5. Asoro FG, Newell MA, Beavis WD, Scott MP, Jannink J-L (2011) Accuracy and training population design for genomic selection on quantitative traits in elite North American oats. Plant Genome 4:132–144
6. Lorenz AJ, Smith KP, Jannink J-L (2012) Potential and optimization of genomic selection for Fusarium Head Blight resistance in six-row barley. Crop Sci 52:1609–1621
7. Hayes BJ, Bowman PJ, Chamberlain AC, Goddard ME (2009) Invited review: genomic selection in dairy cattle: progress and challenges. J Dairy Sci 92:433–443
8. Daetwyler HD, Pong-Wong R, Villanueva B, Woolliams JA (2010) The impact of genetic architecture on genome-wide evaluation methods. Genetics 185:1021–1031
9. Habier D, Fernando R, Dekkers J (2007) The impact of genetic relationship information on genome-assisted breeding values. Genetics 177:2389–2397
10. Habier D, Tetens J, Seefried FR, Lichtner P, Thaller G (2010) The impact of genetic relationship information on genomic breeding values in German Holstein cattle. Genet Sel Evol 42:5
11. Clark SA, Hickey JM, Daetwyler HD, van der Werf JHJ (2012) The importance of information on relatives for the prediction of genomic breeding values and the implications for the makeup of reference data sets in livestock breeding schemes. Genet Sel Evol 44:4
12. de Roos APW, Hayes BJ, Goddard ME (2009) Reliability of genomic breeding values across multiple populations. Genetics 183:1545–1553
13. Hayes BJ, Bowman PJ, Chamberlain AC, Verbyla K, Goddard ME (2009) Accuracy of genomic breeding values in multi-breed dairy cattle populations. Genet Sel Evol 41:51
14. Yu J, Pressoir G, Briggs WH, Bi IV, Yamasaki M, Doebley JF, McMullen MD, Gaut BS, Nielsen DM, Holland JB, Kresovich S, Buckler ES (2006) A unified mixed-model method for association mapping that accounts for multiple levels of relatedness. Nat Genet 38:203–208
15. Jannink J-L (2010) Dynamics of long-term genomic selection. Genet Sel Evol 42:35
16. Bijma P (2012) Accuracies of estimated breeding values from ordinary genetic evaluations do not reflect the correlation between true and estimated breeding values in selected populations. J Anim Breed Genet 129:345–358
17. Sonesson AK, Woolliams JA, Meuwissen THE (2012) Genomic selection requires genomic control of inbreeding. Genet Sel Evol 44:27
18. Heffner EL, Lorenz AJ, Jannink J-L, Sorrells ME (2010) Plant breeding with genomic selection: gain per unit time and cost. Crop Sci 50:1681–1690
19. Endelman J (2011) Ridge regression and other kernels for genomic selection with R package rrBLUP. Plant Genome 4:250–255
20. Wickham H (2009) ggplot2: elegant graphics for data analysis. Springer, New York
21. Lorenz A, Chao S, Asoro F, Heffner E, Hayashi T, Iwata H, Smith K, Sorrells M, Jannink J-L (2011) Genomic selection in plant breeding: knowledge and prospects. In: Sparks DL (ed) Advances in agronomy. Academic, San Diego, CA, pp 77–123

Chapter 11

Simulated Breeding with QU-GENE Graphical User Interface

Adrian Hathorn, Scott Chapman, and Mark Dieters

Abstract

Comparing the efficiencies of breeding methods with field experiments is a costly, long-term process. QU-GENE is a highly flexible genetic and breeding simulation platform capable of simulating the performance of a range of different breeding strategies and for a continuum of genetic models ranging from simple to complex. In this chapter we describe some of the basic mechanics behind the QU-GENE user interface and give a simplified example of how it works.

Key words Marker breeding, Selection schemes, Gene pyramiding

1 Introduction

The ultimate goal of plant breeding programs is to develop new and superior genotypes to those currently available for a specific target environment, or a target population of environments (TPE). In order to achieve this objective, a range of selection methods have been employed by plant breeders. Comparing the efficiencies of these methods with field experiments is a costly, long-term process. Moreover, whilst quantitative genetic theory lays a solid mathematical platform for the design and analysis of selection methods, many of the assumptions in quantitative genetics can seldom, if ever, be tested in reality (e.g., no G×E, no linkage). With computer simulation, many of these simplifying assumptions can be relaxed.

QU-GENE is an integrated genetic and breeding simulation platform capable of simulating the performance of a breeding strategy for a continuum of genetic models ranging from simple to complex [1, 6]. This is made possible through its two stage architecture that separates the definition of the genetic model and genotype environment system (G-E system) from the detail of the individual simulation experiments.

Delphine Fleury and Ryan Whitford (eds.), *Crop Breeding: Methods and Protocols*, Methods in Molecular Biology, vol. 1145, DOI 10.1007/978-1-4939-0446-4_11, © Springer Science+Business Media New York 2014

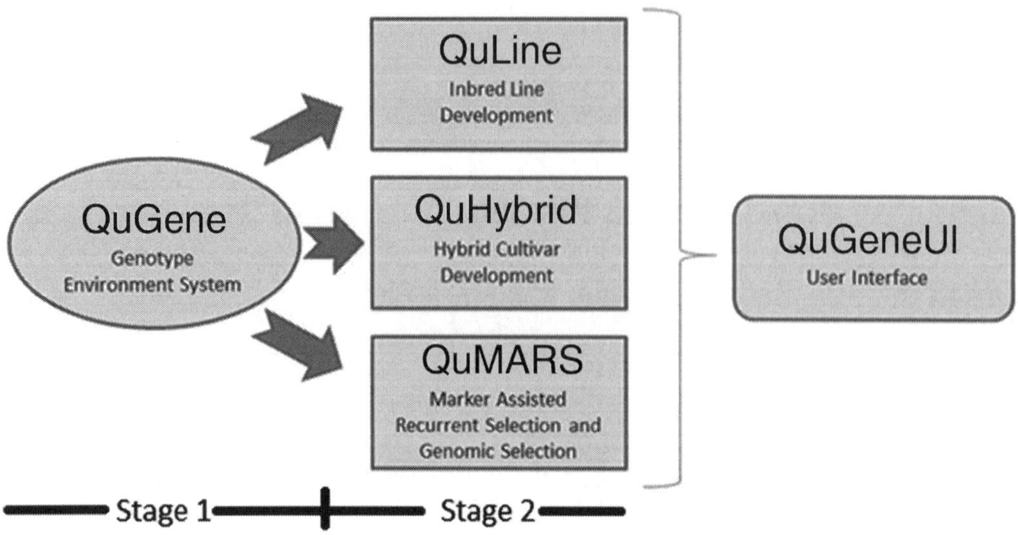

Fig. 1 Application modules for each stage of analysis in QU-GENE

The first stage involves the QU-GENE engine, which has two important functions (Fig. 1). Its first function is to define the G-E system, including all the genetic and environmental information for the simulation experiment. Its second function is to generate the initial population of individuals (base germplasm) so that genetic and/or environmental variances can be estimated. The second stage involves the application modules, whose role it is to investigate, analyze, or manipulate the initial population of individuals within the G-E system as defined by the QU-GENE engine (Fig. 1). The application module usually represents the operation of a breeding program, e.g., [2]. The QU-GENE engine generates core information for the G-E system, whilst the details of each simulation experiment are defined in the application modules.

Using this two stage architecture, numerous alternative simulation experiments that combine breeding methods with a range of G-E systems can be conducted to study the performance of breeding methods under various genetic models, including additive and dominance genetic effects, epistasis, G×E interaction, pleiotropy, linkage, multiple alleles, and molecular markers. Breeding methods that can be simulated include; mass selection, pedigree selection, bulk-population selection, backcross breeding, topcross breeding, doubled haploid breeding, and marker-assisted selection. For a more comprehensive description of this software and the underlying application modules, we recommend reading [1]. Examples of how QU-GENE can be used to simulate realistic breeding scenarios described in [2–4, 15].

2 Installation

QU-GENE is compiled to run in a Windows environment. It has been extensively tested in Windows XP and Windows 7, but still requires further testing in Windows 8. Although the QU-GENE Engine and Application Modules can be run from within a command window using inputs from text files, first time users are advised to install and run QU-GENE with the user interface (QU-GENEUI). More advanced users may choose to run QU-GENE simulations outside of the interface with the assistance of detailed manuals developed for the modules (Qu-Line and Qu-Hybrid) and accessible from the Help function in QU-GENE.

The QU-GENE user interface can be downloaded by going to the QU-GENE homepage (https://www.integratedbreeding.net/supplementary-toolbox/qu-gene) and clicking on "software." Once you have installed the latest version of the QU-GENEUI setup file, you are ready to use QU-GENE.

3 Menu Description

The QU-GENE GUI is based on libraries developed for the APSIM cropping systems simulation tools (www.apsim.info). The top menu has facilities to create, save, and open previous .QUI files which store the information for a set of breeding simulations. Other buttons enable the running of QU-GENE on a given G-E System ("Run QU-GENE"), editing of several folder settings ("Settings"), direct access to Example simulations ("Open Examples"), access to viewing tools for results ("Tools"), and access to help guides for the GUI and the breeding modules.

The "New simulations" menu allows users to drag and drop simulations of parts of simulations from a template to the main window. The "Manage" button allows users to create their own "Toolbox" of simulation parts and drag them down from the main window for use in later simulations.

4 Your Turn

4.1 Starting a New Project

1. Begin by clicking on the "New" tab (Fig. 2).
2. Select "G-E System" as your simulation type and press OK (Fig. 2).
3. You will be asked to select a destination folder on your computer to save the project file (Fig. 2). You can name the project file "example.qui." After saving the project file you should see a new "G-E Systems" node branching off the "Simulations" folder.

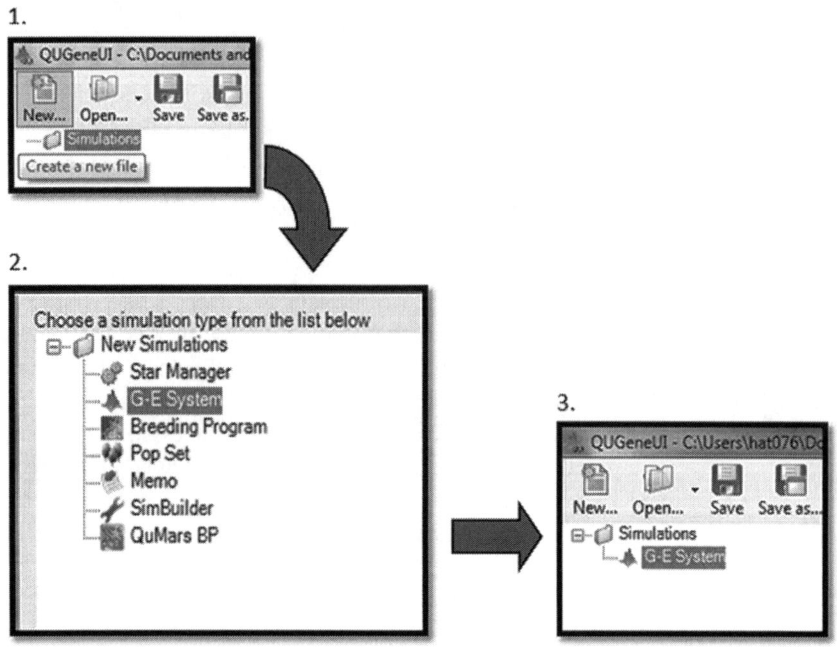

Fig. 2 Screenshots of QU-GENE showing the three steps for starting a new project

4.2 Preparing a New Simulation

1. Open the toolbox by clicking on the "New Simulations" button located at the bottom left hand corner of the screen. The toolbox contains a list of simulation types, or modules, which you can add to your project at any stage.
2. For this basic example we will only add two more modules to the project, a "Breeding Program" and a "Sim Builder." To add a module simply "drag and drop" the required module into the "Simulations" folder. The modules should appear as extra nodes.

4.3 Loading an Example Dataset

1. We will now load the "G-E System" with example data. Begin by clicking on the "G-E System" node in the left side menu. This will bring up a panel to the right of the screen with a number of button options, as well as a display box called "G-E System Summary" (Fig. 3).
2. Click on "Create Example" (Fig. 3).
3. When asked whether you would like to replace the G-E System with an example, click "OK" (Fig. 3). QU-GENE will now load a simple example dataset into the G-E System. You should now see a number of extra sub-nodes available under the "G-E System" (*see* **Note 1**). The default example in QU-GENE (Trait1QTL3) examines a single trait (yield) with three additive genes and six flanking markers, sampled within two environments. This is a simplistic dataset that is intended to

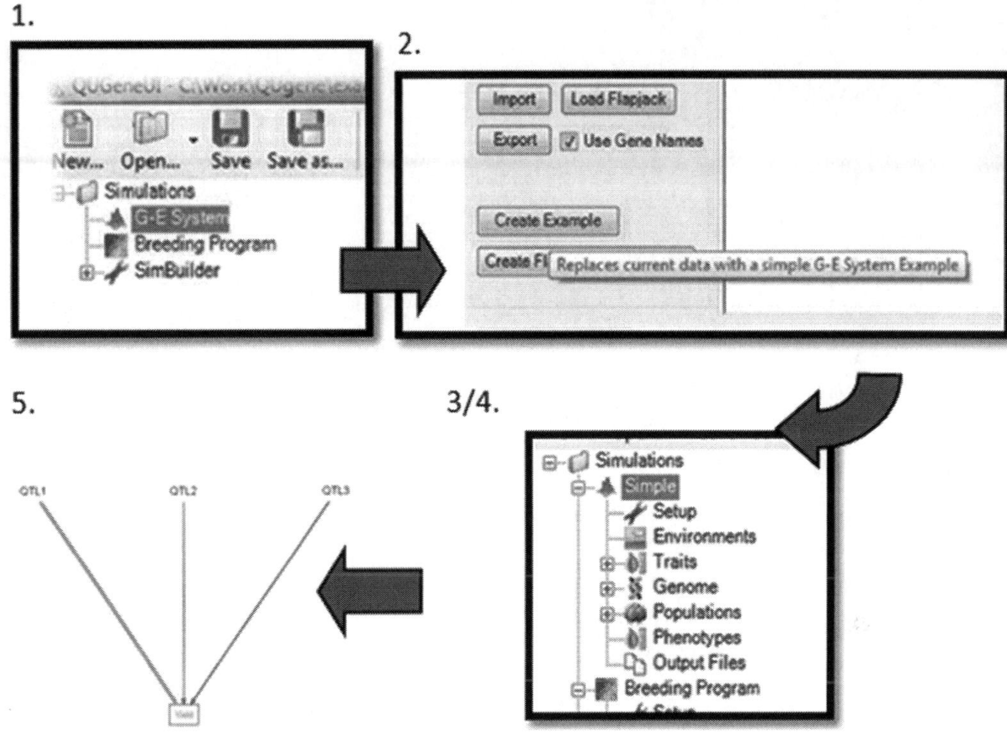

Fig. 3 Screenshots of QU-GENE showing the five steps for loading the example dataset

introduce the novice user to some basic QU-GENE functionality. More complicated example datasets based can be opened by clicking on the "Open Example" tab located on the upper primary tool bar.

4. Now click on the "G-E Systems" node (Fig. 3). You will notice that the "G-E System Summary" box has now been filled and contains a brief description of your simple G-E System. The summary describes a single trait with three QTL effects in two environments, but this could just as easily be a multi-trait analysis with over a 100 genes in multiple environments (*see* **Note 2**).
5. To the right of the summary box is a "View Graph" button. Click on this button to see a graphical depiction of our simple trait—gene network (Fig. 3).

4.4 Loading an Example Breeding Program

1. Click on the "Breeding Program" node.
2. Click on the "Create Example" button then click "OK". Qu-Gene should now load an example breeding program.
3. To view a graphical depiction of the breeding program in landscape format click on "Breeding Program" and then "View Graph" (Fig. 4). We can see that we are comparing two selection strategies, side by side, a low pressure strategy where the

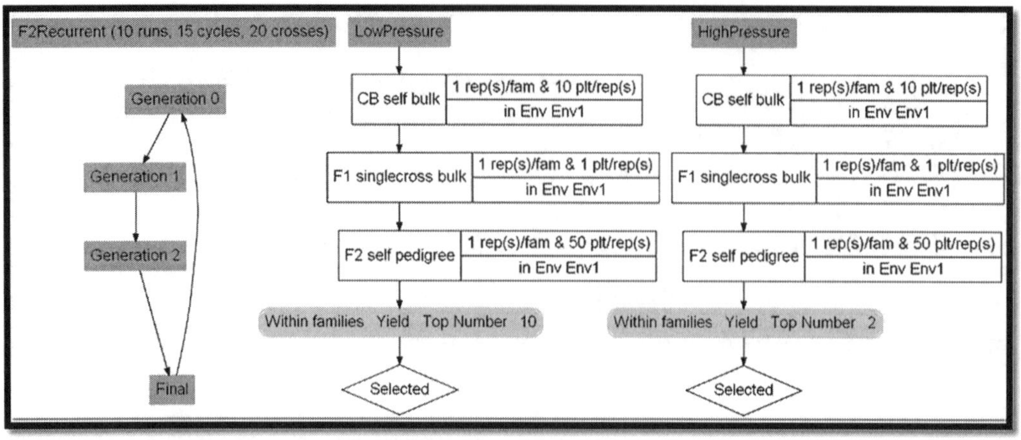

Fig. 4 Graphical depiction of the breeding program in landscape format

top ten individuals (top 25 %) are selected and a high pressure strategy where the top two individuals (top 4 %) are selected. The strategies are identical in other respects. Once the simulation has been run, we will demonstrate how to use QU-GENE to compare the results of these selection strategies using a variety of graphical tools (*see* **Note 3**).

4.5 Running a Simulation

1. Before a simulation can be run, the G-E System(s) must be built and in the process a number of output files need to be generated. On the main toolbar at the top of the screen you will find a button named "Run QU-GENE". Click on it.
2. In the pop up window, check the name of the G-E System you want to run and click "Run Systems." The output files can be seen under "Output Files," a sub-node of the main project node "Simple." You can expand "Output Files" to see these files.
3. Simulations can now be executed by combining the G-E System with the breeding program on the "SimBuilder" node. Click on the "SimBuilder" node.
4. The resulting panel contains all the options available to you for designing your simulation. You must first decide which application module you intend to use (Fig. 5). The default selection is "Qu-Line", but you will notice two other options: "Qu-Hybrid" and "Qu-Mars", "Qu-Hybrid" is used for testcross and hybrid prediction whilst in "Qu-Mars" selection can be based on a prediction of molecular markers in a marker assisted recurrent selection (MARS) or genome wide selection context. All the other options in the "Breeding Programs," "G-E Systems," and "Populations" boxes can be left at their default values.

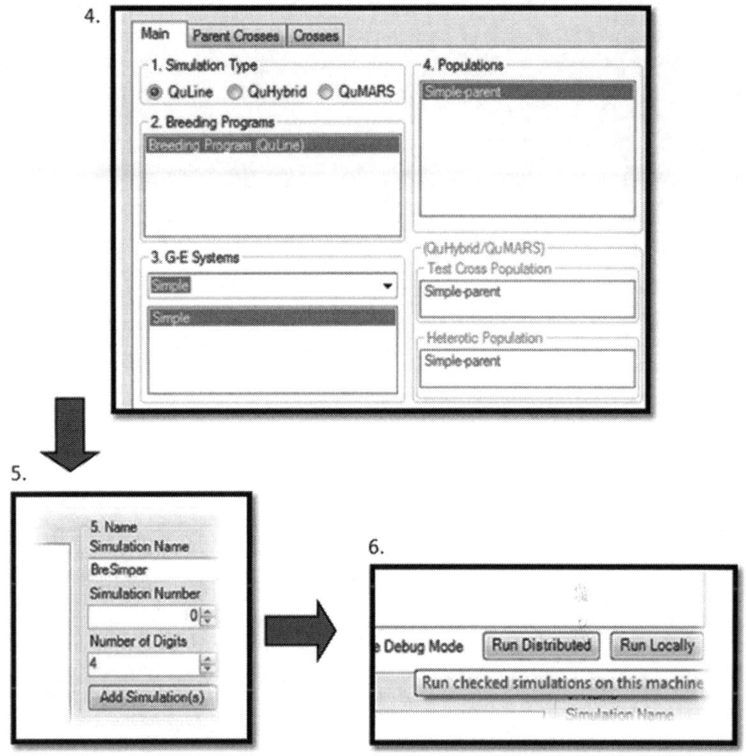

Fig. 5 Screenshots of QU-GENE showing the steps 4, 5 and 6 for running a simulation

5. In the "Name" section you can choose a different simulation name (Fig. 5). Add the simulation to the queue by clicking on "Add Simulation(s)."
6. The simulation should now appear checked under the "Simulations" tab at the top of the screen. Click on "Run Locally" to begin the simulation (Fig. 5).

4.6 Viewing the Results of a Simulation

1. The simulation should only take a few seconds to run. Once it is finished, click on the "results" section, a sub-node of the "Simbuilder." Here you will see a summary of the simulation results and a number of graphical viewing tools.
2. Click on "View Traits."
3. At the bottom of the Trait Viewer there are a number of filtering options which allows the user to observe the effect of selection on trait value in slightly different ways (Fig. 6). To compare both selection strategies side by side click on "Stack Line Chart" (*see* **Note 4**).
4. Highlight both selection strategies in the "Strategy" box. Make sure that all cycles are highlighted, and that we are looking at environment 1 and trait 1.

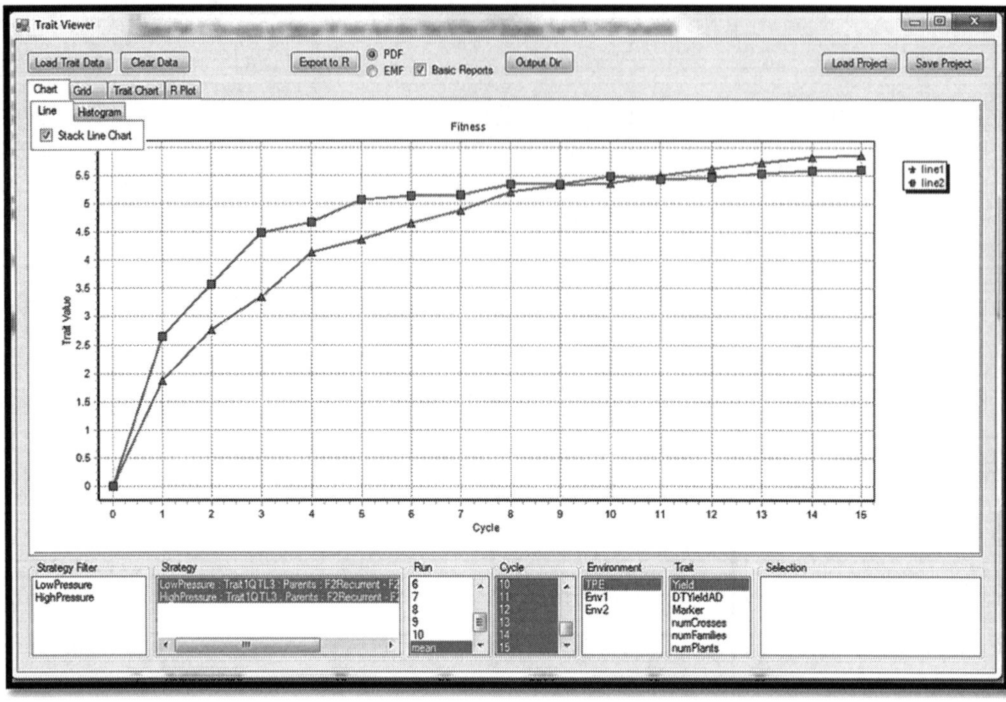

Fig. 6 Trait Viewer window showing the results of a simulation

Notice the effect of selection intensity on the average trait values in the population. When we select the top two individuals, our average trait value increases rapidly as the positive allele at QTL1 (+5) becomes fixed. With a lower selection pressure, fixation of positive alleles takes longer, and thus, the pace of genetic change is reduced.

5. Now open up the Landscape Viewer by clicking on "View Landscape" (Fig. 7) (*see* **Note 4**).
6. Highlight both selection strategies as well as cycle 0. The frequency of the positive allele should begin at 0.5.
7. Once cycle 0 has been highlighted, move through the remaining cycle numbers using the down arrow key. Notice that the frequency of the positive allele increases at different rates for both strategies. By cycle 8, QTL 1 has been virtually fixed under both strategies although this can be seen to occur more rapidly under a high selection pressure.

5 Advanced Simulation Examples

The Qu-GeneUI is supplied with a set of examples based on published research in the manipulation of genes and traits using different breeding strategies. Each example contains a memo tab which

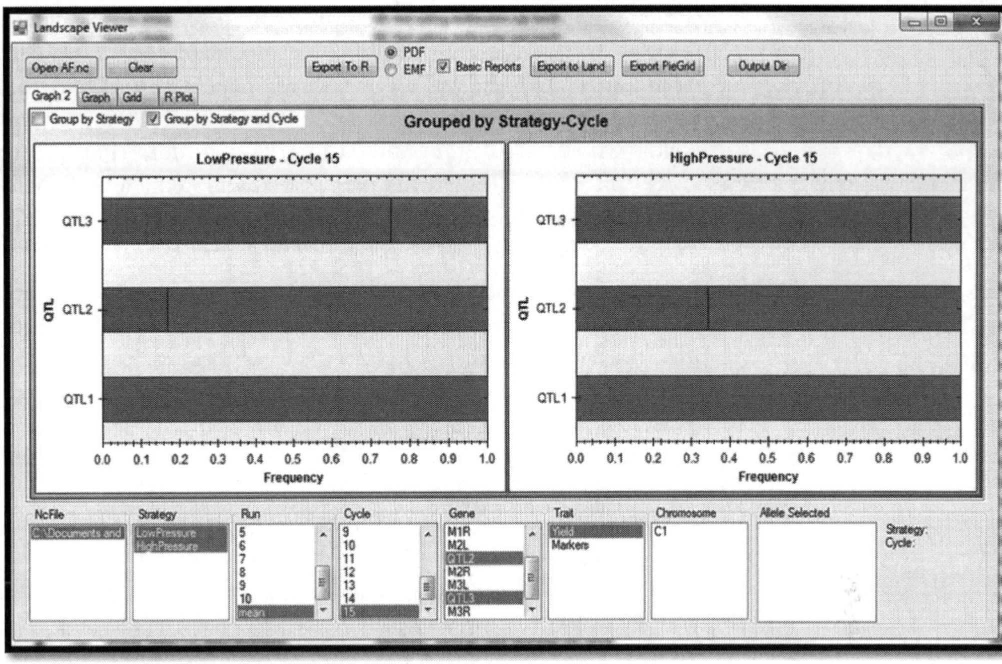

Fig. 7 Landscape Viewer window showing the results of a simulation

contains a detailed description of the example. These include the following examples:

1. CIMMYTWheat—Comparing long-term strategies to improve wheat using complex trait and gene models derived from knowledge and experience of breeders. The objective of this research was to conduct a simulation experiment in which the engine QU-GENE and the application module Qu-Line were used to compare CIMMYT's MODPED and SELBLK methods [10] in terms of genetic gain, number of crosses retained after one breeding cycle, and resource allocation [12, 13]. The experiment evaluates long-term selection gains for these breeding programs, using complex genetic models to explain gene action and trait heritability, based on knowledge and experience of CIMMYT breeders.
2. GluteninModel—Example of gene pyramiding by crossing Australian wheat lines, applying single-seed descent and the applying phenotypic screening for two quality traits while varying the order and intensity of selection [2, 3, 14].
3. MaizeF3_9env—The objective was to conduct a simulation using results of an extensive QTL analysis of performance of maize F3 lines grown in nine environments. The original research [4, 5] estimated QTL effects in test-crosses (i.e. additive effects) for grain yield and grain moisture. These analyses were repeated by Malosetti et al. (unpublished 2011), with

similar QTL estimated, but including some QTL × Environment effects. In this example, the QTL results of Malosetti are used to initialize the GES with broad sense heritability of the traits set either to 100 % or to typical observed values (50 % for grain yield, 80 % for grain moisture). An index trait (grain yield = 2.5 × grain moisture) was also calculated. Selection for grain yield, grain moisture or the index was undertaken with 100 % (perfect marker) or reduced heritability.

4. Qu-MARS—The objective of this research was to conduct a simulation experiment in which the engine QU-GENE and the application module Qu-MARS were used to compare different numbers of cycles (one, two or three) of genomic selection (based on BLUP from all markers in a "QTL" experiment). The gene and trait data (for grain yield and moisture) are those for additive QTL identified from the dataset of [7]. See also the other Qu-GeneUI example "F3_9env." The breeding program design is based on the sorghum user case (Jean-Francois Rami, unpublished) being developed within the GCP IBP. The BLUP estimation used in Qu-MARS is that of Bernardo and Yu [1]. The results show that after two cycles of GS, the accumulation of major QTL associated with grain yield has been achieved for the first environment (E201). This program is designed to be extended with a large number of small QTL effects for further testing.
5. RiceMABCsub1—Introgression of the sub1 (submergence tolerance) QTL in rice. The purpose of this example was to demonstrate that the fixation of favorable QTL, [8, 9], can be achieved with fewer individuals per replicate in the BC2F1 and BC2F2 populations.
6. WheatColeoptile—combining major genes and QTL for wheat coleoptile length and plant height using combinations of marker and phenotypic selection. The Qu-Line genetic and breeding simulation tool was used to investigate a realistic wheat breeding scenario where MAS for major genes and QTL of smaller effects are integrated with conventional phenotypic selection to develop new lines from an elite biparental cross. The breeding objective was to combine six major reduced-height, disease resistance, and grain quality genes with phenotypic selection and MAS to maximize coleoptile length in target breeding lines [10, 16].

6 Notes

1. Building the QU-GENE engine input file (.qug) from scratch is a complicated process. Should you want to build your own GE-System, it is recommended that you load the example

dataset and then adjust the example GE-System within an Excel spreadsheet. To do this, click on the "Save XLS" button, choose your destination folder and begin making changes. Once you have made the changes, load the new G-E System into QU-GENE by clicking on the "Load XLS" button.

2. Many of the nodes, such as "Traits" and "Genome" can also be expanded to provide more detailed information about your G-E System. For example, in "Genome" → "Gene Map" → "Gene Effects" you will notice a table containing a list of all the QTL in the model and their effect values in each environment type. Notice QTL1 has a positive effect value of +5 in both environments; QTL2 has a negative effect value of −1 in both environments whilst QTL3 has a small positive effect of +1 in environment type 1 and a negative effect of −2 in environment type 2.
3. Before running a simulation, note that there are other nodes that can be added to a simulation: "Memo" which is a notepad to store formatted notes about a simulation, and "Pop Set" which can be used as a place-holder for "populations" that are used in later simulations.
4. In both the trait and landscape viewers, there are options to create more complex PDF reports of the results by using the "R Plot" tab. This requires "R" (www.r-project.org) to be installed on the computer.

References

1. Bernardo R, Yu J (2007) Prospects for genome-wide selection for quantitative traits in maize. Crop Sci 47:1082–1090
2. Eagles HA, Hollamby GJ, Gororo NN, Eastwood RF (2002) Estimation and utilization of glutenin gene effects from the analysis of unbalanced data from wheat breeding programs. Aust J Agr Res 53:367–377
3. Eagles HA, Eastwood RF, Hollamby GJ, Martin EM, Cornish GB (2005) Revision of the estimates of glutenin gene effects at the Glu-B1 locus form southern Australian wheat breeding programs. Aust J Agr Res 55:1093–1096
4. Moreau L, Monod H, Charcosset A, Gallais A (1999) Marker assisted selection with spatial analysis of unreplicated field trials. Theor Appl Genet 98:234–242
5. Moreau L, Charcosset A, Gallais A (2004) Use of trial clustering to study QTL × environment effects for grain yield and related traits in maize. Theor Appl Genet 110:92–105
6. Podlich DW, Cooper M (1998) QU-GENE: a simulation platform for quantitative analysis of genetic models. Bioinformatics 14:632–653
7. Huang Y, Madur D, Combes V, Long Ky C, Coubriche D, Jamin P, Jouanne S, Dumas F, Bouty E, Bertin P, Charcosset A Moreau L (2010) The genetic architecture of grain yield and related traits in Zea maize L. revealed by comparing intermated and conventional populations. Genetics 186:395–404
8. Prigge V, Maurer HP, Mackill DJ, Melchinger AE, Frisch M (2008) Comparison of the observed with the simulated distributions of the parental genome contribution in two marker-assisted backcross programs in rice. Theor Appl Genet 116:739–744
9. Prigge V, Melchinger AE, Dhillon BS, Frische M (2009) Efficiency gain of marker-assisted backcrossing by sequentially increasing marker densities over generations. Theor Appl Genet 119:23–32
10. Rebetzke GJ, Ellis MH, Bonnett DG, Richards RA (2007) Molecular mapping of genes for coleoptile growth in bread wheat (Triticum aestivum L.). Theor Appl Genet 114:1173–1183
11. van Ginkel M, Trethowan R, Ammar K, Wang J, Lillemo M (2002) Guide to bread wheat

breeding at CIMMYT (rev), Wheat special report no. 5. CIMMYT, Mexico, D.F., Mexico

12. Wang J, van Ginkel M, Podlich D, Ye G, Trethowan R, Pfeiffer W, DeLacy IH, Cooper M, Rajaram S (2003) Comparison of two breeding strategies by computer simulation. Crop Sci 43:1764–1773
13. Wang J, van Ginkel M, Trethowan R, Ye G, DeLacy I, Podlich D, Cooper M (2004) Simulating the effects of dominance and epistasis on selection response in the CIMMYT wheat breeding program using QuCim. Crop Sci 44:2006
14. Wang J, Eagles HA, Trethowan R, van Ginkel M (2005) Using computer simulation of the selection process and known gene information to assist in parental selection in wheat quality breeding. Aust J Agric Res 56:465–473
15. Wang J, Chapman SC, Bonnett DG, Rebetzke GJ, Crouch J (2007) Application of population genetic theory and simulation models to efficiently pyramid multiple genes via marker-assisted selection. Crop Sci 47:582
16. Wang J, Chapman S, Bonnett D, Rebetzke G (2009) Simultaneous selection of major and minor genes: use of QTL to increase selection efficiency of coleoptile length of wheat (*Triticum aestivum* L.). Theor Appl Genet 119:65–74

Chapter 12

The Control of Recombination in Wheat by *Ph1* and Its Use in Breeding

Graham Moore

Abstract

Two meiotic processes have a major influence on the plant breeding, namely, the independent assortment of chromosomes, and recombination. The major chromosome pairing locus in hexaploid and tetraploid wheat, *Ph1*, has a significant effect on both these processes. This chapter reviews our current understanding of this locus and how mutants of it can be exploited for breeding purposes.

Key words Meiosis, Chromosome, Crossover, Condensation, Polyploid

1 Introduction

Breeders wishing to exploit the potential of wild relatives carrying beneficial traits in their own breeding programs ultimately rely on the ability of the chromosomes from the crop and the wild species carrying the desired trait to undergo crossover formation during meiosis. These breeders exploit interspecific crosses between wheat and its wild relatives to introduce useful agronomic traits (for example disease resistance) into wheat. Generally it is possible to successfully generate such interspecific hybrids containing a haploid set of wild relative and wheat chromosomes. Occasionally for some species such as rye, a barrier (the crossability locus) exists which prevents successful hybridization. However, the major barrier in the exploitation of such interspecific hybrid in wheat is the *Ph1* locus. This locus when present successfully suppresses recombination between the wheat and wild relative chromosomes during meiosis, thereby blocking this introduction. Mutants of *Ph1*, which carry a deletion of this locus, have been exploited in wide-crossing programs to allow the wild relative and wheat chromosomes to exchange [1]. Recent studies have increased our understanding of

Delphine Fleury and Ryan Whitford (eds.), *Crop Breeding: Methods and Protocols*, Methods in Molecular Biology, vol. 1145, DOI 10.1007/978-1-4939-0446-4_12,

the mode of action of this locus, and therefore what avenues might be pursued to better manipulate this process. This chapter reviews the progress that has been made thus far.

2 Meiosis: The Basic Process

Meiosis is a cell division process that ensures gametes carry the correct number of chromosomes, without a doubling of chromosome number. It is therefore central to the life cycle of all sexual eukaryotic organisms. As a result of its being an ancient process, both the recombination and segregation mechanisms are largely conserved across most eukaryotic organisms. Organisms exhibiting sexual reproduction carry two copies (homologues) of each chromosome. Before meiosis, each homologue is replicated, forming two sister chromatids that remain linked together. At the start of meiosis, each chromosome must recognize its homologue from among all of the chromosomes present in the nucleus. The homologues must then become intimately aligned or paired along their entire lengths. A proteinaceous structure known as the synaptonemal complex (SC) is assembled between them in a process termed synapsis (reviewed, 2). Meiotic recombination (the exchange of DNA strands) is completed in the synapsed structure, resulting in the crossover formation between the DNA strands of the homologues or parental chromosomes, thereby reshuffling genetic information. Also importantly meiotic recombination forms chiasmata, physical links which together with sister chromatid cohesion hold the homologues together after the disassembly of the SC. This enables homologues to be correctly orientated on the first meiotic spindle. The homologues are then segregated during meiosis I so that each gamete carries only a single copy of each chromosome. A second division occurs in which the sister chromatids are segregated so that each gamete carries only a single copy of each chromosome [3].

3 Breeding Depends on Recombination, So What Is the Distribution of Recombination?

The manipulation of the meiotic processes could have a profound impact on commercial plant breeding. Crossovers created during meiotic recombination not only generate new combinations of alleles, but, in most organisms, are essential to ensure the correct segregation of the homologues to gametes, and hence high fertility of a species [3]. Therefore, it is not surprising that both the number and distribution of crossovers along chromosomes are tightly controlled in most organisms including crop plants. Clearly the barriers that affect or modify the ability to recombine specific chromosome regions are a crucial issue in breeding. So what are the barriers?

There are at least three constraints that affect crossover distribution. Firstly, each chromosome pair must possess at least one crossover, this is known as the obligatory crossover. Secondly, crossovers are not independent of each other, the presence of a crossover inhibits the formation of a new crossover nearby on the chromosome and this inhibition decreases with distance. This process is known as crossover interference [4]. Thirdly, crossovers are not uniformly distributed along chromosomes; some regions, known as hotspots, display much higher frequencies of crossing over than others. This last point is very important for breeding programs as it implies that the distribution of crossovers with respect to the location of genes will affect the efficiency of the breeding process. Many plants with large genomes (for example wheat, barley, Lolium, Festuca, and rye) exhibit an uneven distribution of recombination along their chromosomes with the proximal chromosome regions exhibiting reduced recombinational activity [5]. Recombination has been shown to occur mostly in genes, and in particular in their promoter regions. For many large genomed plants, their genomic sequence is not yet available so it is unclear whether the regions lacking recombinational activity also lack genes. However, comparative genome analysis does provide us with an indication of the gene content of the proximal chromosomal regions. Early studies revealed that whole rice chromosomes (particularly linkage groups 8, 10, 11 and 12) are equivalent to the proximal regions of chromosomes in wheat, rye, barley, and Lolium [6, 7]. These rice chromosomes contain thousands of genes and therefore by implication, so must the equivalent regions in these species. Studies exploiting barley carrying chromosomal deletions also indicated that rice linkage groups could be allocated to cytologically defined regions [8, 9]. Thus, significant numbers of genes are likely to be present in these regions with significantly reduced recombinational activity. Recent studies exploiting comparative mapping in Lolium/Festuca, wheat and rice provide further support for this proposal [7]. Thus, in the future, it is going to be clearly important in breeding strategies for these large genomed species to be able to alter the distribution of recombination so that recombination could be increased in regions exhibiting extensive suppression. At the moment, there is no published strategy for addressing this issue in wheat.

4 What Are the Sites of Recombination?

Most of the intervening DNA between genes in the large genomes is composed of repetitive elements or retroelements. Therefore, it might be thought that the presence of highly similar repetitive sequences on all the chromosomes might complicate the pairing and recombination between chromosomes. However, comparative sequencing of cereals and grasses revealed that although the genes are largely conserved in equivalent regions in related species, the repetitive elements are not conserved. Sequence comparisons of the repetitive content of wheat chromosomes with their diploid progenitors also revealed a lack of conservation of repeat sequences in the intervening regions suggesting a rapid turnover [10]. In the most extreme case of maize, the intervening regions between two varieties can be markedly different [11]. Thus, recombination between the chromosomes of related species or in the case of maize, between the chromosomes of two varieties is going to be confined largely to the homologous regions or the genic regions [11]. A recent study in wheat has confirmed that recombination is largely confined to the genic regions [12].

5 What Are the Barriers to Recombination Between Chromosomes in Breeding?

A major barrier in manipulating recombination is the fact that more than 70 % of flowering plants are polyploid [13]. Polyploids arise as a result of either doubling of chromosomes following sexual hybridization within the same species (autopolyploid) or between closely related species containing related but not completely homologous (homoeologous) genomes (allopolyploids). Some of the world's most important crops are allopolyploids, including wheat, canola, oats, cotton, tobacco, cultivated sugarcane, and cooking banana. Some plants, such as maize, are cryptic allopolyploids, that is, following polyploidization, genome rearrangements have now partially concealed their polyploid origin. Allopolyploids must behave as diploids during meiosis to produce viable gametes and be fertile. Thus, homologues need to be distinguished from related chromosomes (homoeologues) with similar gene content and order but divergent in repetitive DNA content. The overall efficiency and accuracy of the mechanisms used to achieve this diploidization have a profound influence on the fertility of every allopolyploid plant, and this is clearly of major importance for the success of a crop species in which seed is the economic product.

Many wild species carry interesting characteristics that would be useful to exploit in breeding programs. So what are the barriers to their exploitation? Clearly the ability to produce interspecific hybrids is an important issue. However, sexual hybridization between a polyploid and a wild species generally produces an interspecific hybrid containing a haploid set of polyploid and wild-relative chromosomes. The exploitation of wild relative variation during breeding of wheat for example benefited from an understanding of the genome relationships of wheat and its wild diploid and polyploid relatives. Early studies by Kihara and colleagues had established many of the genomic relationships of the diploid and polyploid *Triticum* and *Aegilops* species on the basis of chromosome pairing in hybrids.

However, in many cases, these analyses revealed that there is a low level of pairing and recombination between crop and wild-relative chromosomes. The factors that affect the ability to pair and recombine these chromosomes are often those involved in controlling chromosome pairing between the related chromosomes within the polyploid itself. Therefore, an understanding the basis for this pairing suppression has a practical effect in being able to be switch it on and off, thereby enhancing breeding strategies [14].

6 Chromosome Pairing and Recombination in Wheat and Its Interspecific Hybrids

The hexaploid wheat (AABBDD, $2n=42$) carries a diploid set of 42 chromosomes composed of three ancestral genomes, A, B, and D, while tetraploid wheat (AABB, $2n=28$) carries a diploid set of 28 chromosomes composed of two ancestral genomes. The 42 chromosomes can be divided into seven groups of six chromosomes (two chromosomes from each of the ancestral genomes). In hexaploid wheat, chromosome 1A must pair with 1A at meiosis and not with 1B or 1D. Both Riley and Sears observed that the loss of the 5B chromosome resulted in pairing between the related wheat chromosomes [15–17]. The region when deleted causing the effect on related chromosome pairing is termed the *Ph1* locus (*P*airing *h*omoeologous *1*) (Fig. 1). However, other loci that affect pairing in wheat have also been identified including *Ph2* on 3DS as well as loci on chromosomes 5AL, 5DL, 5AS, 3AL, 3BL, and 3DL [18]. A locus suppressing the activity of *Ph1* action has also been identified [19]. To date, no other loci have been identified which compensate for the absence of the *Ph1* locus in hexaploid and tetraploid wheat. In comparison to the characterization of other pairing loci, *Ph1* has been the most extensively studied to date. Some information is available on *Ph2* in wheat which appears to affect chromosome synapsis [20]. A number of candidate genes for *Ph2* have been

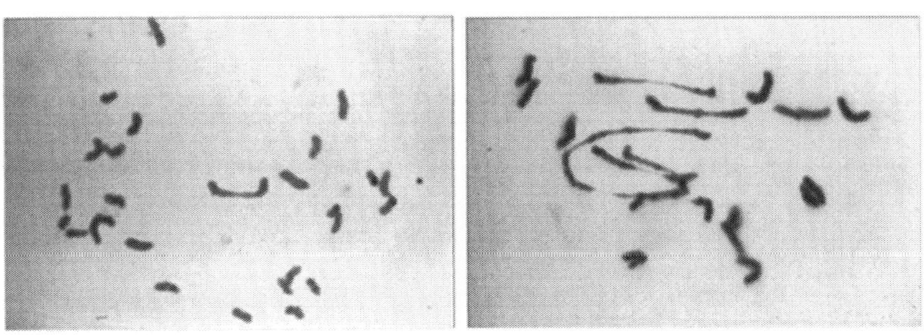

Wheat-rye hybrid carrying the *Ph1* locus   Wheat-rye hybrid lacking the *Ph1* locus

Fig. 1 Chromosome pairing during meiosis in wheat–rye hybrids at metaphase I

linked to this region which may have a role in this process based on rice and wheat comparative mapping [21, 22].

Chromosome pairing in tetraploid and hexaploid wheat during meiosis is largely restricted to homologous chromosomes in the presence of the *Ph1* locus on chromosome 5B. The other homoeologous chromosomes 5A and 5D do not compensate for loss of its activity. The 5G chromosome from the wheat relative *T. timopheevi*, when substituted for 5B does provide a level compensation [23]. *T. timopheevi* is likely to have arisen from wide crossing between the same progenitors which gave rise to tetraploid wheat 0.5 million years earlier.

Deletion of the *Ph1* locus in hexaploid wheat produces a subtle meiotic phenotype. Chromosomes pair and recombine correctly in half the cells (or meiocytes) progressing through meiosis. In the remaining meiocytes, there is a low level of pairing failure and pairing between multiple chromosomes. With each generation of the mutant, this disruption in pairing increases due to accumulated chromosome rearrangements and eventually results in the mutant being infertile [24–26]. Thus, *Ph1* stabilizes wheat's polyploid genome, resulting in its high fertility. However, the *Ph1* locus has a more extreme effect on meiotic chromosome pairing in interspecific hybrids between wheat and wild relatives. These hybrids possess a haploid complement of chromosomes and therefore only related chromosomes can pair at meiosis as there are no homologues. The presence of *Ph1* prevents any recombination from occurring between the related chromosomes and so the chromosomes are observed as univalents at metaphase I. In the absence of *Ph1*, recombination occurs and a number of chromosomes are paired at metaphase I [27]. Thus, *Ph1* suppresses recombination between related chromosomes in these hybrids. Given the subtle effect of *Ph1* in wheat compared with its effect in hybrids, many researchers have exploited wheat hybrids to score for the presence and absence of *Ph1*.

Ph1 alters the initiation of chromosome condensation during meiosis. Thus, the related chromosomes are in different condensation states when mispaired associations between chromosomes are resolved into crossovers or not ([28, 29], Martinez et al. unpublished). Okadaic acid treatment at the onset of meiosis induces premature condensation, and hence chromosome pairing between related chromosomes in hybrids. This should eliminate having to use the *Ph1* mutant line. However, the problem is accessibility of the drug to the meiocytes. If this can be solved, using an approach where viable seed could be recovered, then would improve the efficiency of the process. *Ph1* also affects independent centromere pairing and therefore chromosome segregation, and hence assortment [30], as well as homologue recognition [31]. At the initiation of meiosis, homologues undergo a synchronized elongation process before they intimately pair [31]. This process is disrupted when *Ph1* is deleted leading to an increase in incorrect associations. The chromatin elongation process connected with homologue recognition has now been shown in *C. elegans* and *Tetrahymena*, and independent centromere pairing has now been shown in *Arabidopsis*, *Brachypodium*, and *Tetrahymena* suggesting that these processes first reported in wheat are general cross-species biological phenomena. However, although *Ph1*'s control of centromere pairing and homologue recognition is important for stabilizing correct pairing, assortment, and segregation of chromosomes during meiosis in wheat itself, it is not so important for inducing pairing and recombination between related chromosomes for breeding. *Ph1*'s effect on meiotic condensation is both relevant to correct chromosome pairing and recombination in wheat itself, and to inducing recombination between related chromosomes.

7 Exploitation of Chromosome Pairing Loci

This topic has been extensively reviewed for example by Qi et al. [32]. Sears transferred disease resistance loci from wild relatives of wheat by recombining the chromosomes with those of wheat in lines *Ph1* deficient [24]. More recently CIMMYT exploited *Ph1* to introgress from closely related species [33]. However, over the years with retirements, the long-term nature of wide-crossing programs and the difficulty of identifying introgressed segments from wild relatives in wheat itself apart from cytogenetically, most wide crossing programs have closed.

However, the UK has just re-established a wheat pre-breeding program exploiting the *ph1* mutant line in 2011 under Professor Ian King, University of Nottingham. The major difference between this wide-crossing program now and those in the past is the availability of high-throughput genotyping platforms which can rapidly identify from which chromosome alien introgressed segment is derived, its size, and its location in the wheat genome. This changes the efficiency of such wide-crossing programs. A full description of the program of research in BBSRC's wide crossing program is available at http://www.wheatisp.org. The primary objective of this wide crossing program is to transfer small alien chromosome segments, carrying target genes but lacking deleterious genes, into wheat quickly and efficiently. The strategy is essentially as developed by Sears, i.e., the identification and intercrossing of lines that carry different and overlapping chromosome segments followed by the analysis of the resulting progeny to isolate small introgressions. This is achieved in two ways: (1) transfer of the entire genome of alien species into wheat in overlapping chromosome segments and (2) targeted development of overlapping chromosome segments for specific regions of the genome carrying target genes. Four diploid ($2n=2x=14$) species have been selected for this work: *T. urartu*,

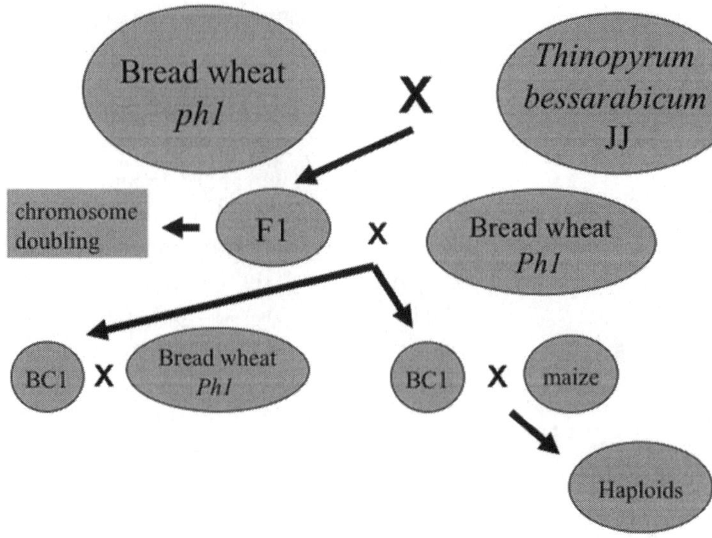

Fig. 2 Strategies for exploiting the *Ph1* mutant in wheat breeding programs

Th. bessarabicum, rye, and *Aegilops speltoides*. The strategy deployed rapidly yields genome wide introgressions in the BC_1 in only two generations.

In order to induce wheat–alien homoeologous recombination, related species are first crossed to *Ph1* mutant lines, i.e., lines that lack the *Ph1* locus (see summary of Strategies, Fig. 2). Homoeologous recombination will occur at meiosis in the F_1. In the UK program, the *Ph1* deletion is in Paragon (Pa) background. Paragon is an elite UK variety, more suited to trait analysis and is widely crossable to a range of wild relative species. The new *Ph1* deletion in Pa background is available from Graham Moore (John Innes Centre). Thus, the four interspecific F_1 are then crossed with normal Pa carrying the wild type *Ph1* locus to generate BC_1 populations that will contain wheat–alien introgressions. More than 1,000 BC_1 progeny need to be generated from each of the four wheat–alien F_1 populations, i.e., 4,000 progeny in total.

Crossing wheat involves two stages: first emasculation and then pollination. Emasculation precedes controlled pollination. Emasculations are carried out as the ear emerges from the flag leaf, roughly 3–7 days before natural self-pollination occurs. The top 1/3rd and bottom 1/3rd portions of florets along a wheat ear of the Paragon *Ph1* mutant are clipped off using scissors or forceps, leaving only the middle spikelets. Then the middle florets in these spikelets are removed, leaving only the two primary florets (lateral), which are used for the next step of the emasculation and pollination. Spikelets are then cut and opened exposing the insides of the florets enabling the three anthers present in each spikelet to be removed using forceps. Care must be taken during emasculation not to damage the gynoecium which will be pollinated by the male donor. To prevent contamination from pollen, the emasculated spikelets are bagged.

Three to seven days after emasculation, the gynoecium is ready for fertilization. The anthers from the desired wild relative parent are taken and inserted in an inverted position onto the gynoecium of the "female flower" in order to disperse the pollen and thereby pollinate the flower. After ensuring the abundance of pollen, the plastic bag was closed again (to prevent cross-pollination), and the ear was labeled with the name of the female plant (+the date of the emasculation) × the name of the male plant used for pollination and the date of the pollination. Embryo rescue is sometimes required after wide crossing, if the endosperm of the hybrid seed does not develop. Also colchicine treatment has also been exploited to rescue recombinant chromosomes from wheat–wild relative crosses by doubling of the chromosomes. So for example a strategy employed in crossing the wild relative *Thinopyrum bessarabicum* onto the *ph1* mutant can be as follows. The resulting F1 wheat–*Thinopyrum* hybrid is backcrossed onto wild type wheat. The F_2 progeny are then crossed with maize and the resulting progeny colchicine treated to double the chromosomes (*see* Chapter 16). The protocol used as JIC for undertaking this is described below, which is essentially producing double haploid lines by intergeneric crossing with maize.

Alien introgressed chromosome segments are identified exploiting a range of genotyping platforms (*see* Chapters 2–5). This (1) confirms the presence of an introgression, (2) reveals the physical size of an introgression, and (3) determines the position of recombinant events. This will enable comparisons to be made between marker profiles and the size of alien segments introgressed and also to determine the presence/absence of the physical clustering of wheat–alien recombinant events.

By exploiting different genotyping platforms, rapid detection of most, if not all, alien introgressions generated in the BC_1 population (both single and double wheat–alien recombinants in a chromosome arm) can be made. Once identified, selected introgressions are characterized at high resolution genotyping using high-throughput SNP platforms to determine (1) the exact size of each alien chromosome segment and (2) to identify lines carrying overlapping chromosome segments with a high degree of precision. Once characterized, each introgression line is backcrossed once more to wheat to remove the remaining *Ph1* deletion locus and then self-fertilized to produce plants homozygous for the alien introgression. Introgression homozygotes and the genomic insertion point of alien chromosome segments can be all determined. These plants will be self-fertilized to bulk up seed that will be provided to breeders and researchers for trait analysis. When a specific region of the alien genome is identified as carrying a specific target gene, the overlapping introgressions for that region can be identified and intercrossed, resulting in yet smaller introgressions carrying a target gene can be identified.

References

1. Sears E (1977) An induced mutant with homoeologous pairing in common wheat. Can J Genet Cytol 19:585–593
2. Zickler D, Kleckner N (1999) Meiotic chromosomes: integrating structures. Annu Rev Genet 33:603–754
3. Page SL, Hawley RS (2003) Chromosome choreography: the meiotic ballet. Science 301: 785–789
4. Jones GH (1984) The control of chiasma distribution. Symp Soc Exp Biol 38:293–320
5. Akhunov E, Goodyear A, Geng S, Qi L-L, Echalier B, Gill B, Miftahudin P, Gustafson P, Lazo G, Chao S, Anderson O, Linkiewicz M, Dubcovsky J, Rota M, Sorrells M, Zhang D, Nguyen H, Kalavacharla V, Hossain K, Kianian S, Peng J, Lapitan N, Gonzalez-Hernandez J, Anderson J, Choi D-W, Close T, Dilbirligi M, Gill K, Walker-Simmons K, Steber C, McGuire P, Qualset C, Dvorak J (2003) The organisation and rate of evolution of wheat genomes are correlated with recombination rates along the chromosome arms. Genome Res 13:753–763
6. Moore G (1995) Cereal genome evolution: pastoral pursuits with 'Lego' genomes. Curr Opin Genet Dev 5:717–724
7. King J, Armstead A, Donnison A, Roberts L, Harper J, Skot K, Elborough K, King I (2007) Comparative analyses between Lolium/Festuca introgression lines and rice reveal the major fraction of functionally annotated gene models are located in recombination poor/very poor regions of the genome. Genetics 177:547–606
8. Korzun L, Kunzel G (1996) The physical relationship of barley chromosome 5 (1H) to the linkage group of rice chromosome 5 and 10. Mol Gen Genet 252:225–231
9. Kunzel G, Korzun L, Meister A (2000) Cytogenetically integrated physical restriction length polymorphism maps of the barley genome based on translocation breakpoints. Genetics 154:397–412
10. Scherrer B, Isidore E, Klein P, Kim J-S, Bellec A, Chalhoub B, Keller B, Feuillet C (2005) Large intraspecific haplotype variability at the Rph7 locus from rapid and recent divergence in the barley genome. Plant Cell 17:361–374
11. Fu H, Dooner H (2002) Intraspecific violation of genetic colinearity and its implications in maize. Proc Natl Acad Sci U S A 99:9573–9578
12. Saintenac C, Favre S, Remay C, Choulet F, Ravel C, Paux E, Balfourier F, Feuillet C, Sourdille P (2011) Variation in crossover rate across a 3-MB contig of bread wheat (*Triticum aestivum*) reveals the presence of a meiotic recombination hotspot. Chromosoma 120:185–198
13. Materson J (2003) Stomatal size in fossil plants: evidence of polyploidy in majority of Angiosperms. Science 264:421–424
14. Able J, Langridge P (2006) Wild sex in grasses. Trends Plant Sci 11:261–267
15. Okamoto M (1957) A synaptic effect of chromosome V. Wheat Inf Serv 5:6
16. Riley R, Chapman V (1958) Genetic control of the cytologically diploid behaviour of hexaploid wheat. Nature 182:713–715
17. Sears ER, Okamoto M (1958) Intergenomic chromosome relationships in hexaploid wheat. Proc Xth Int Cong Genet Montreal 2:258–259
18. Feldman M (1993) Mode of action of *Ph1* in wheat. Crop Sci 33:894–897
19. Dvorak J, Deal KR, Luo MC (2006) Discovery and mapping of wheat *Ph1* suppressor. Genetics 174:17–27
20. Martinez M, Cuñado N, Carcelén N, Romero C (2001) The *Ph1* and *Ph2* loci play different roles in the specific behaviour of hexaploid wheat *Triticum aestivum*. Theor Appl Genet 102:751–758
21. Dong C, Whitford R, Langridge P (2002) A DNA mismatch repair gene links to the *Ph2* locus in wheat. Genome 45:116–124
22. Sutton T, Whitford R, Baurnann U, Dong M, Able J, Langridge P (2003) The *Ph2* pairing homoeologous locus in wheat (*Triticum aestivum*): identification of candidate meiotic genes using a comparative genetics approach. Plant J 36:443–456
23. Ozkan H, Feldman M (2001) Genotypic variation in tetraploid wheat affecting homoelogous pairing in hybrids with *Aegilops peregrine*. Genome 44:1000–1006
24. Sears E (1972) Agropyron-wheat transfers through induced homoeologous pairing. Can J Genet Cytol 14:746
25. Roberts MA, Reader SM, Dalgliesh C, Miller TE, Foote TN, Fish LJ, Snape SW, Moore G (1999) Induction and characterisation of *Ph1* wheat mutants. Genetics 153:1909–1918

26. Sanchez-Moran E, Benavonte E, Orellana J (2001) Analysis of karyotypic stability of homoeologous-pairing mutants in allopolyploid wheat. Chromosoma 110:371–377
27. Dhalinal HS, Gill BS, Waines JG (1977) Analysis of induced homoeologous pairing in a *ph* mutant wheat x rye hybrid. J Hered 68:206–209
28. Knight E, Greer E, Draeger T, Thole V, Reader S, Shaw P, Moore G (2010) Inducing chromosome pairing through premature condensation: analysis of wheat interspecific hybrids. Funct Integr Genomics 10:603–608
29. Greer E, Martinez-Ramirez A, Pendle A, Colas I, Jones AEM, Moore G, Shaw P (2012) The *Ph1* locus suppresses Cdk2-type activity during premeiosis and meiosis in wheat. Plant Cell 24:152–162
30. Martinez-Perez E, Shaw P, Moore G (2001) The *Ph1* locus is needed to ensure specific somatic and meiotic centromere association. Nature 411:204–207
31. Prieto P, Shaw P, Moore G (2004) Homologue recognition during meiosis is associated with change in chromatin conformation. Nat Cell Biol 6:906–908
32. Qi L, Friebe B, Zhang P, Gill BS (2007) Homoeologous recombination, chromosome engineering and crop improvement. Chromosome Res 15:3–19
33. William HM, Singh RP, Trethowan R, Van Ginkel M, Pellegrinshi A, Huerta-Espin A, Hosington D (2005) Biotechnology applications for wheat improvement CIMMYT. Turk J Agric 29:113–119

Chapter 13

TILLING for Plant Breeding

Peter Sharp and Chongmei Dong

Abstract

TILLING is widely used in plant functional genomics. Mutagenesis and SNP detection is combined to allow for the isolation of mutations in genes of interest. It can also be used as a plant breeding tool, whereby variation in known or candidate genes of interest to breeding programs is generated. Here we describe a simple low-cost TILLING procedure.

Key words CEL 1, Mismatch, Mutation, SNP

1 Introduction

TILLING (*T*argeting *I*nduced *L*ocal *L*esions *IN* *G*enomes) is a technique that combines two well-known technologies, mutagenesis of plants by chemical mutagens, coupled with high-throughput screening for single nucleotide polymorphisms (SNP) by mismatch detection. This combination was first described in 2000 and elaborated on in subsequent studies [1–4]. The technique is widely used for plant functional genomics [2, 3], and can also be used in "prebreeding" [4], which is the discovery or generation of genetic variation of potential benefit to commercial and public plant breeding programs that release crop cultivars to farmers.

Various technologies can be used in TILLING-like research—different mutagens and mutation detection methods [4–8].

Most mutations in an EMS mutagenized population are single nucleotide mutations randomly distributed throughout the genome. To identify mutations in a gene of interest it is necessary to be able to detect single nucleotide polymorphisms (SNP) between the mutant and wild type. A popular TILLING technology to identify SNP is based on the use of a mismatch-specific nuclease from celery, CEL 1. In this protocol, pooled DNA is amplified using fluorescently labelled gene-specific primers (with different fluorophores on the 3′ and 5′ tags). The amplified products are denatured by heating and then cooled slowly allowing

Delphine Fleury and Ryan Whitford (eds.), *Crop Breeding: Methods and Protocols*, Methods in Molecular Biology, vol. 1145, DOI 10.1007/978-1-4939-0446-4_13,

random re-annealing. Heteroduplex molecules form when a pool includes at least one plant with a mutation in the amplified fragment. The re-annealed products are digested with CEL 1 nuclease, which cleaves the heteroduplexed mismatch site. Cleaved fragments, which are resolved on gels (polyacrylamide denaturing gel such as a LI-COR gel system, or agarose gel), identify pools, and therefore individual samples that have a mutation within the amplicon. The position of the mutation can be estimated based on the size of the fragments carrying 3′ and 5′ fluorescent tags [3].

This chapter describes a modified TILLING procedure which is fast for establishing a mutagenized population and low-cost for screening of mutations. The method presented here has been used in bread wheat (*Triticum aestivum*) but could be applied to other crops [7]. This protocol uses crude celery juice extract of CEL 1 [9], unlabeled primers, and agarose gels for mutation detection.

2 Materials

Prepare all solutions using ultrapure water (prepared by purifying deionized water to attain a sensitivity of 18 MΩ cm at 25 °C). Prepare and store all reagents at room temperature (unless indicated otherwise). Follow all waste disposal regulations when disposing of waste materials.

2.1 Mutagenesis

1. Seeds of chosen cultivar (*see* **Note 1**).
2. Glasshouse facility which can accommodate at least two hundred 10 cm pots. Field facilities are needed for large-scale plant propagation (i.e., growing populations of a couple of thousand individuals).
3. EMS solution: EMS (ethyl methanesulfonate, Sigma Catalogue No: M0880) is made freshly to 0.5 % (v/v) in water (*see* **Note 2**), using a magnetic stirrer to completely dissolve the EMS.
4. Seed storage room with controlled temperature and humidity.
5. Cheese cloth.

2.2 DNA Preparation

1. DNA extraction buffer: 0.1 M Tris–HCl (pH 8.0), 0.05 M EDTA (pH 8.0), and 1.25 % SDS, autoclave before use.
2. 6 M ammonium acetate: autoclaved and stored at 4 °C.
3. TE/RNaseA solution: 10 mM Tris–HCl and 1 mM EDTA (pH 8.0), autoclaved. Add RNaseA to TE to get 50 μg/mL before use. Leftover solution can be stored at 4 °C.
4. 100 % isopropanol.
5. 70 % ethanol.
6. Agarose.

7. GelRed (Biotium Inc., Hayward, CA, USA) or Ethidium Bromide (Sigma)
8. GoTaq Hot Start Polymerase kit (Promega, Madison, WI, USA).

2.3 CEL 1 Nuclease Preparation

All procedures should be performed at 4 °C.

1. Weigh ~500 g celery sticks (freshly purchased from a supermarket). Make juice using a juice maker.
2. To 400 mL of juice, add 44 mL 10× buffer (1 M Tris–HCl, pH 7.7) and 440 μL 200 mM PMSF (Phenylmethanesulfonyl fluoride, dissolved in 100 % ethanol), and mix with a magnetic stirrer.
3. Divide the juice into eight 50 mL centrifuge tubes and spin at 2,600 × *g* for 20 min.
4. Combine supernatants in a beaker to get 300 mL, add 43.2 g $(NH_4)_2SO_4$ to make 25 % (add slowly while mixing). Mix for 30 min.
5. Centrifuge at 15,000 × *g* for 40 min.
6. Get 300 mL supernatant, add 117 g $(NH_4)_2SO_4$ to make 80 % (add slowly while mixing). Mix for 30 min.
7. Spin at 16,000 × *g* for 1.5 h. Pour off the supernatant.
8. Dissolve the pellet in 1× buffer (100 mM Tris–HCl, pH 7.7, 100 μM PMSF). Add 2 mL of buffer for each tube (total eight tubes). Dissolve overnight at 4 °C.
9. Dialyze using 12,000 D cutoff tube in 10 L 1× buffer. Change the buffer every 1–2 h for a total of four changes.
10. Carefully pour the dialyzed juice extract into a clean 50 mL falcon tube. Add the same volume of glycerol to the juice extract to make this crude enzyme to 50 % glycerol. Aliquot 100–500 μL of the enzyme in 1 mL eppendorf tubes and keep at −20 °C, or for longer term storage, use −80 °C. The enzyme is now ready to be used as CEL 1.

2.4 DNA Screening

1. CEL 1 as prepared in Subheading 2.3.
2. 2× CEL 1 digestion buffer: 20 mM HEPES (4-(2-hydroxyethyl)-1-piperazine-ethanesulfonic acid) pH 7.5, 20 mM $MgSO_4$, 20 mM KCl, 0.004 % TritonX-100, and 0.4 μg/mL BSA (bovine serum albumin), which is sterilized by passing through a 0.22 μm filter. Store the buffer at −20 °C.
3. 6× loading buffer: 10 mM Tris–HCl, pH 8.0, 60 mM EDTA, 0.03 % bromophenol blue, 0.03 % xylene cyanol FF, 60 % glycerol.
4. PCR machine.
5. Benchtop microcentrifuge (such as Eppendorf 5430 and Eppendorf 5810R).

6. QIAGEN Mixer Mill (Retsch mill, Type MM 300).
7. Laboratory oven and water bath.
8. 4 mm stainless steel ball bearing.
9. Plastic ware such as PCR plate, eppendorf tubes, 1.1 mL 8-strip tubes and 8-strip caps and BulkRacks (National Scientific Supply, California, USA), 0.8 mL storage plate (ABgene).
10. Agarose gel electrophoresis equipment.
11. GelDoc-It imaging system (UVP LLC, Upland, CA, USA).
12. Silica gels for drying leaf tissue.
13. 8-channel 1 mL, 200 μL, 10 μL pipettes, 12-channel 10 μL pipette.

2.5 Software Programs for Primer Design and Sequence Analysis

1. CODDLE (http://www.proweb.org/coddle/).
2. Primer 3 (http://frodo.wi.mit.edu/primer3/).
3. BioEdit (http://www.mbio.ncsu.edu/bioedit/bioedit.html).
4. Mutation Surveyor® (SoftGenetics, State College, PA, USA).
5. PARSESNP (http://www.proweb.org/parsesnp/).

3 Methods

3.1 Preparation of the Mutant Population

1. Mutagenesis

Approximately 2,000–3,000 seeds are divided into six portions with each portion wrapped in cheese cloth then soaked in 1 L of 0.5 % EMS solution (a ratio of about 300 seeds per 100 mL EMS solution) in a 2 L beaker (*see* **Notes 2** and **3**). The beaker is then put on a shaker securely and shaking at a low speed (about 50–60 rpm) for 18–20 h (overnight). After the EMS treatment, seeds are taken out of the solution (*see* **Note 4**) and then extensively washed under running tap water for at least 2 h before sowing in 10 cm pots (about 20 seeds per pot). Pots are placed in a glasshouse (18 °C) for 3 weeks to allow germination. These seedlings, termed the M1 generation, should then be transplanted to field.

2. Seed bank and DNA sampling

For a resource limited project, the following method is suitable for the harvest and storage of seeds. Given that each M1 plant can produce multiple spikes, each spike may contain different mutations. In order to fully exploit the population, each spike derived from a single M1 plant can be harvested separately. For example, if three spikes are harvested from a single plant, the seeds will be labelled as SEED1.1, SEED1.2 and SEED1.3 in three seed bags. For the second plant, seeds will be labelled as SEED2.1, SEED2.2

and SEED2.3, and so on. These seeds should be stored in a seed storage room in a numerical order to become a seed bank. For DNA sampling, a single seed from each seed bag should be germinated in 100-cell germination trays to give (M2) seedlings; this usually takes two weeks to process. After leaf tissue is sampled for DNA extraction, seedlings can be discarded.

Another method is single seed descent (SSD) which can prevent multiple sampling of the same mutation. This can occur when more than one spike per M1 plant is taken. A single seed from each M1 plant should be harvested then grown to produce M2 plants. Leaf tissue of M2 plants should then be collected for DNA extraction. Harvest seeds of individual M2 plants and store seeds at 4 °C under low humidity. This material represents the seed bank. This method requires a large M1 population, significant plant growth facilities, labor and time for growing M2 plants to maturity.

3.2 DNA Extraction

High quality DNA can be extracted from the leaf tissue by any preferred method, either using a commercial kit such as the QIAGEN plate-formatted kit or a noncommercial method such as CTAB or SDS. The following method is a modified SDS method adopted for using 8-strip tubes in a 96-well format; use 8-channel pipette for liquid handling in all steps.

1. Collect the leaf sample (about 6 cm long) in a 1.1 mL 8-strip tube, leave it open and dry the leaf tissue in a Silica Gel box with tightly closed lid for about 5 days. Change to fresh silica gel once.
2. Place two 4 mm ball bearing into each tube and then close it with 8-strip caps. Place the 8-strip into the BulkRack so that 96 samples can be processed at once.
3. Grind the dried leaf samples (in 96-format) in the QIAGEN Mixer Mill for 2 min at the speed of 22 oscillations/s, repeat once (or until leaf tissue becomes fine powder).
4. Add 600 μL of preheated (65°C) DNA extraction buffer to each tube, briefly vortex and incubate at 65 °C (water bath) for 15–30 min (*see* **Note 5**). Shake the samples once during incubation.
5. Allow the samples to cool to room temperature and add 300 μL of 6 M ammonium acetate which is stored at 4 °C. Shake vigorously to mix-in the ammonium acetate and leave to stand for 15–30 min in a 4 °C fridge.
6. Centrifuge the samples for 30 min at the max speed (4,000 rpm/3,220 rcf in Eppendorf 5810R, *see* **Note 6**), pipette out 700 μL supernatant into new 8-strip tubes.
7. Add 420 μL of isopropanol to each sample and mix thoroughly by inversion. Allow the DNA to precipitate for 5 min at room temperature.

8. Pellet the DNA by centrifuging the samples for 10 min at the max speed in Eppendorf 5810R, and then pour off the supernatant by inverting the 8-strip tubes.
9. Add 500 μL of 70 % ethanol to each sample for washing, stand at room temperature for 10–20 min, and then centrifuge at max speed for 5 min. Pour off the ethanol solution, then spin again for 2 min.
10. Carefully remove the 70 % ethanol wash solution by pipetting out. Dry the DNA pellet on the bench for about 30 min.
11. Suspend the pellet in 200 μL TE/RNase solution. Leave the DNA to dissolve overnight at 4 °C in the fridge. Or the DNA can be dissolved at room temperature or 37 °C oven.
12. Briefly vortex the sample, centrifuge the sample at max speed for 20 min. There may be an un-dissolved pellet, which is not DNA.
13. Transfer the supernatant to a 96-well storage plate. Check the DNA concentration by loading 1–2 μL on an agarose gel together with a known concentration λ DNA for comparison; finally adjust DNA to ~10–15 ng/μL for PCR.

3.3 DNA Pooling

After DNA samples are quantified and normalized to a uniform concentration, DNA samples should be pooled for TILLING screening. Classic TILLING methods use eightfold pooling [3], here we use a fourfold, 2-way pooling strategy.

1. Using a 12-channel pipette transfer DNA from A1-A12, B1-B12, C1-C12, and D1-D12, so that they fill only the first row (A-row) of a new 96-well plate. This represents the pooled DNA samples in the first A-row, i.e., sample A1 of the new plate contains a mix of samples (A1, B1, C1, D1) from the original plate; sample A2 of the new plate contains a mix of samples (A2, B2, C2, and D2) from the original plate; and so on (Fig. 1).
2. Using only four channels of an 8-channel pipette, transfer DNA from A1 to D1, A2–D2, A3–D3, and A4–D4 of the old plate to the second row filling B1 to B4 of the new plate, continue transferring DNA from A5 to D5, A6–D6, A7–D7, A8–D8 of the old plate to B5–B8 in the second row of the new plate, likewise A9–D9, A10–D10, A11–D11, A12–D12 of the old plate to B9–B12 in row two of the new plate. The resultant new 96-well plate will be a fourfold, 2-way DNA pool (Fig. 1).

These 2-way pooled DNA samples are used as template for gene specific amplification by PCR and subsequent CEL 1 digestion of the PCR products. Pooled samples allow for screening multiple DNA samples at once for mutations in a desired gene fragment (Fig. 2). The presence of each individual DNA sample in two separate pools allows the easy identification of those individual plant

	1	2	3	4	5	6	7	8	9	10	11	12
A	A1 B1 C1 D1	A2 B2 C2 D2	A3 B3 C3 D3	A4 B4 C4 D4	A5 B5 C5 D5	A6 B6 C6 D6	A7 B7 C7 D7	A8 B8 C8 D8	A9 B9 C9 D9	A10 B10 C10 D10	A11 B11 C11 D11	A12 B12 C12 D12
B	A1 A2 A3 A4	B1 B2 B3 B4	C1 C2 C3 C4	D1 D2 D3 D4	A5 A6 A7 A8	B5 B6 B7 B8	C5 C6 C7 C8	D5 D6 D7 D8	A9 A10 A11 A12	B9 B10 B11 B12	C9 C10 C11 C12	D9 D10 D11 D12
C	E1 F1 G1 H1	E2 F2 G2 H2	E3 F3 G3 H3	E4 F4 G4 H4	E5 F5 G5 H5	E6 F6 G6 H6	E7 F7 G7 H7	E8 F8 G8 H8	E9 F9 G9 H9	E10 F10 G10 H10	E11 F11 G11 H11	E12 F12 G12 H12
D	E1 E2 E3 E4	F1 F2 F3 F4	G1 G2 G3 G4	H1 H2 H3 H4	E5 E6 E7 E8	F5 F6 F7 F8	G5 G6 G7 G8	H5 H6 H7 H8	E9 E10 E11 E12	F9 F10 F11 F12	G9 G10 G11 G12	H9 H10 H11 H12
E												
F												
G												
H												

Fig. 1 A 96-well plate showing a fourfold, 2-way pooling strategy for DNA samples derived from an original 96-well DNA sample plate

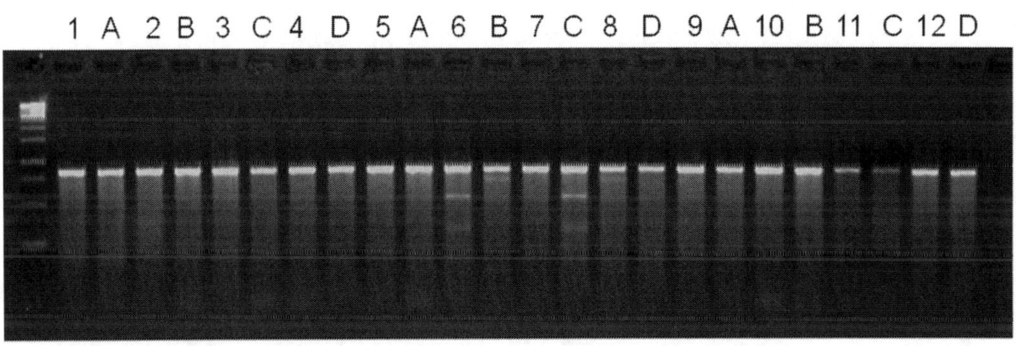

Fig. 2 Samples in the A-row and B-row in Fig. 1 after PCR and CEL 1 digestion were loaded on an agarose gel by a 12-channel pipette. After electrophoresis, from the CEL 1 digestion pattern, two mutants, C6 and B8, were identified

lines containing mutations. Figure 2 is an example of a single mutant line which shows duplicate CEL 1 digest patterns of electrophoretically separated PCR amplicons derived from separate DNA pools. Here, well coordinates for these two pooled samples allows the identification of corresponding coordinates for DNA in the original 96-well plate format and hence the subsequent identification of the mutant line. Furthermore, such a pooling strategy facilitates high-throughput (12-channel pipette) PCR sample preparation, digestion and loading for electrophoresis, therefore saving time, particularly because second PCR and digest screens are not required, as would be the case with one-way pooling.

3.4 Choice of Target Genes and PCR Primer Design

1. Target gene sequence

For breeding purposes, a target gene is typically a gene with a known function. For example, wheat granule-bound starch synthase I (GBSSI) gene is involved in amylose synthesis in wheat grain endosperm. The first successful demonstration of a wheat TILLING population screen allowed the identification of loss-of-function mutants in all three homoeologs, which when combined knocked-out *GBSSI* resulting in waxy wheat (zero-amylose starch) [8]. When a gene of interest is first selected for TILLING, consideration must be made to whether sufficient sequence information is available. With fast growing genomic technologies, a large amount of gene sequence is available in public databases (for example, NCBI http://www.ncbi.nlm.nih.gov/). For wheat many partial cDNA sequences are found at NCBI, and corresponding genomic sequences can be identified from repositories like the UK 454 genomic assemblies derived from the reference cultivar Chinese Spring (http://www.cerealsdb.uk.net/) or the International Wheat Genome Sequencing Consortium (http://www.wheatgenome.org/). Gene structure (predicted exon/intron positions) can be determined by comparison to orthologous genes from a reference genome such as rice and maize, if present.

2. Likelihood of mutation

Once the target sequence has been identified, it can be analyzed for the likelihood of mutation by using the Web-based software CODDLE (http://www.proweb.org/coddle/). CODDLE provides a link for searching homologous protein sequences (Blocks Family). This compares protein sequence of interest against known databases for similarity, therefore identifying functional domains. Regions up to 1.5 kb in size can be identified by CODDLE to have highest likelihood of producing mutations that will deliver likely loss-of-gene function. This prediction is based on the number of nonsense codons (premature stop codon), alteration of a transcript splicing sites, and nonconservative missense mutations in a highly conserved domain. PCR primers can then be designed by Primer3. This software is linked to CODDLE.

The ploidy level of the species must be considered. For example, wheat is an allohexaploid, which contains three closely related genomes. Most "single copy" genes usually have one gene copy on each homoeologous chromosome (A, B and D). Genome specific primers are necessary for TILLING as potential mutations can be confounded by homoeologous SNP. It is therefore necessary to compare homoeologous gene sequences so as to identify the most variable regions between the three gene copies. This is usually within introns where typical site for primer design. Software BioEdit can be used for sequence analysis and multiple sequence comparison.

3.5 Mutation Screening

1. PCR amplification, denaturation, and re-annealing PCR products

PCR amplification of a gene of interest is performed according to the requirement of the primers' annealing temperatures and extension time required by the particular *Taq* polymerase used. DNA polymerase with high fidelity is highly recommended for TILLING.

After PCR, products are then denatured and re-annealed. The procedure can be set on a PCR machine as follow: 95 °C for 5 min, 90 °C for 1 min, then 99 cycles of 89 °C for 30 s with each cycle reducing by 0.5 °C. The PCR products are now ready for CEL 1 digestion.

2. CEL 1 digestion

After the PCR products are denatured and re-annealed, mix 9 μL of PCR product with 10 μL of 2× CEL 1 buffer and 1 μL of CEL 1 (*see* **Note 7**). This step should be performed on ice. Samples are then incubated at 45 °C for 30 min. Stop the digestion by adding 4 μL of 6× loading buffer. The samples are now ready for electrophoretic separation in an agarose gel.

3. Agarose gel electrophoresis for mutation detection

A 2 % agarose gel is made for electrophoresis of digested PCR products. The thinner the gel is, the better the sensitivity for detecting the CEL 1 digested bands. Usually gel thickness is ≤4 mm, with the distance between the bottom of comb teeth and the gel holder being ≤1 mm. High throughput is achieved using combs and gel tanks chosen for compatibility with 12-channel pipetting equipment which accommodates 96-well plate based samples. The DNA staining dye (GelRed or ethidium bromide) can be added in a gel. After electrophoresis of 10 μL samples for 30 min, the gel is visualized in a GelDoc-It imaging system. Gel images are analyzed for the presence of cleavage PCR products using Adobe Photoshop software (Adobe Systems Inc., San Jose, CA). Individual mutant lines can be identified by the CEL 1 digestion patterns and following the pooling strategy as outlined in Fig. 2.

4. Sequencing

Once individual mutants are identified, PCR on the same gene fragment(s) are performed using the individual DNA samples as opposed to the pooled ones. PCR products should be purified for Sanger sequencing. Depending upon the amount of sequencing data to be analyzed, the software Mutation Surveyor® (SoftGenetics, State College, PA, USA) can be employed for the automatic detection of single nucleotide mutations [10]. Once a nucleotide change is identified, it can be further characterized in terms of whether the mutation is a missense, silent, truncation, or splice site mutation.

A computer program PARSESNP (http://www.proweb.org/parsesnp/) is a useful tool for predicting whether the mutation will have a severe impact on protein function.

The last step of using TILLING as a breeding tool is to cross any potentially useful mutant with a wild-type plant for the purpose of modifying the desired trait(s) [7]. Residual mutation load can be removed by repeated backcrosses to wild-type and can be reduced to a minimum after only a few rounds of crossing.

4 Notes

1. Seeds of high genetic uniformity, such as highly homozygous, inbreeding species, are the best choice for creating TILLING population.
2. EMS is a widely used mutagen due to its efficiency in generating a high density of mutations. It is also a well-studied mutagen which causes alkylation of guanine to make G to A and C to T transitions [11]. A lethality test of EMS on the chosen species and seed lot is highly recommended before starting any large-scale mutagenesis. In order to determine the appropriate EMS dosage, a dilution series can be used for treatment on a small amount of seed using the same protocol as stated in Subheading 3.1, but on a smaller scale. Germination rates are determined after two weeks in the glasshouse. A killing curve (germination rate vs. EMS concentration) can then be plotted. The EMS concentration that results in a germination rate of between 65 and 80 % can be used for generating a large-scale mutagenesis population, particularly for species like wheat [7].
3. When seeds are to be wrapped in a cheese cloth, some spare space should be retained as the seeds will expand after imbibing EMS solution.
4. EMS is highly toxic and mutagenic. Handling, storage and disposal must be according to the regulatory framework of your organization.
5. To avoid any cross contamination between samples, add 4-layer paper towel on top of tubes and place BulkRack lid on top of the paper towel, close the lip tightly by using rubber bands to hold the lid and sample tubes (in the BulkRack) together.
6. If using a centrifuge other than Eppendorf 5810R, one can use a higher speed (e.g., 5,000 rcf) which will give a better result.
7. The activity of CEL 1 in the crude extract of celery can differ from batch to batch. It is highly recommended to test the activity of any freshly prepared batch first. Mix a known mutant DNA with wild-type DNA, PCR-amplify the fragment containing the mutation, denature and re-anneal the products,

digest the heteroduplex products with a concentration series of CEL 1 enzyme (for example 0.5, 0.75, 1.0, 1.5, 2.0 μL), and then run a gel. This will allow to identify the most suitable concentration of CEL 1 to be used.

References

1. McCallum CM, Comai L, Greene EA, Henikoff S (2000) Targeted screening for induced mutations. Nat Biotechnol 18:455–457
2. McCallum CM, Comai L, Greene EA, Henikoff S (2000) Targeting induced local lesions in genomes (TILLING) for plant functional genomics. Plant Physiol 123:439–442
3. Colbert T, Till BJ, Tompa R, Reynolds S, Steine MN, Yeung AT, McCallum CM, Comai L, Henikoff S (2001) High-throughput screening for induced point mutations. Plant Physiol 126:480–484
4. Wang T, Uauy C, Till B, Liu C-M (2010) TILLING and associated technologies. J Integr Plant Biol 52:2017–2030
5. Uauy C, Paraiso F, Colasuonno P, Tran RK, Tsai H, Berardi S, Comai L, Dubcovsky J (2009) A modified TILLING approach to detect induced mutations in tetraploid and hexaploid wheat. BMC Plant Biol. doi:10.1186/1471-2229-9-115
6. Dong C, Vincent K, Sharp P (2009) Simultaneous mutation detection of three homoeologous genes in wheat by High Resolution Melting analysis and Mutation Surveyor®. BMC Plant Biol. doi:10.1186/1471-2229-9-143
7. Dong C, Dalton-Morgan C, Vincent K, Sharp P (2009) A modified TILLING method for wheat breeding. Plant Genome 2:39–47
8. Slade AJ, Fuerstenberg SI, Loeffler D, Steine MN, Facciotti D (2005) A reverse genetic, nontransgenic approach to wheat crop improvement by TILLING. Nat Biotechnol 23:75–81
9. Till BJ, Burtner C, Comai L, Henikoff S (2004) Mismatch cleavage by single-strand specific nucleases. Nucleic Acids Res 32:2632–2641
10. Dong C, Yu B (2011) Mutation surveyor: an *in silico* tool for sequencing analysis. In: Yu B, Hinchcliffe MJ (eds) Methods in molecular biology: *in silico* tools for gene discovery. Humana, Totowa, NJ, doi: 10.1007/978-1-61779-176-5_14
11. Greene EA, Codomo CA, Taylor NE, Henikoff JG, Till BJ, Reynolds SH, Enns LC, Burtner C, Johnson JE, Odden AR (2003) Spectrum of chemically induced mutations from a large-scale reverse-genetic screen in *Arabidopsis*. Genetics 164:731–740

Chapter 14

In vitro Culture for Doubled Haploids: Tools for Molecular Breeding

Sue Broughton, Parminder K. Sidhu, and Philip A. Davies

Abstract

Doubled haploid (DH) techniques using immature microspores or embryos rescued in vitro following interspecific hybridization are important tools for breeding wheat and barley and for generating populations for gene mapping purposes. Here, we describe methods being used for the successful production of spring wheat and barley DHs in Australia.

Key words Doubled haploid, Microspore, Embryo rescue, Wheat, Barley, Method

Abbreviations

BA 6-Benzyladenine
IAA Indole-3-acetic acid
2,4-D 2,4-Dichlorophenoxyacetic acid

1 Introduction

Haploid plants are derived from male or female gametic cells. Duplication of the haploid genome, either spontaneously or with chemical treatment, generates "doubled haploids" which are fertile, homozygous plants with significant value for crop breeding and gene mapping [1, 2]. Achievement of homozygosity in a single generation can save 3–5 years in the time required to breed self-pollinating crop varieties and provides additional advantages by increasing the proportion of desirable segregants and improving the accuracy and reliability of phenotypic evaluation [3]. Australian cereal varieties produced by DH breeding include the barley varieties Tantangara, Dhow, Sloop SA, Flagship, Navigator, VT Admiral, Skipper and wheat varieties Hume, Gregory, Gladius, Axe, Crusader, Spitfire, Cobra, Gauntlet, Merlin, Fang, and Espada.

Delphine Fleury and Ryan Whitford (eds.), *Crop Breeding: Methods and Protocols*, Methods in Molecular Biology, vol. 1145, DOI 10.1007/978-1-4939-0446-4_14,

Doubled haploid populations also remain a key research tool for genetic mapping and transformation [4, 5]. Although different types of segregating populations can be used for constructing genetic linkage maps, DH populations offer a number of advantages. Their homozygosity means that they can be multiplied for replicated trials over locations and years, and that populations can be transferred between different laboratories for further linkage analysis and the addition of markers to existing maps [6]. Doubled haploid populations have been widely used in Australian programs to map single locus genes and QTL for a wide variety of traits in wheat and barley [7, 8].

From a molecular breeding perspective, combining DH technology with marker-assisted selection is a genetically effective and economically viable strategy. The design of marker-assisted breeding strategies has been examined by a number of authors, often with the option of using either DH or Single Seed Descent (SSD) populations [9–11]. Marker-assisted screening can either be implemented before DH lines are produced (pre-screening) and/or afterwards (post-screening). Pre-screening can be highly effective when segregating donor plants (e.g., F_2 or F_1 from complex crosses, topcrosses, or backcrosses) are screened with one or more markers and DHs are only generated from donor plants containing the gene/s of interest [3]. This can be complemented by post-screening regenerant plants, where plants are rapidly screened and eliminated at the seedling stage saving time and resources in further grow-outs. Additionally, with plants derived from wheat × maize wide crossing, it is possible to screen and discard haploid plants prior to colchicine application with further time and resource savings.

A variety of methods are used for cereal DH production in Australia. Androgenic methods are preferred for barley (isolated microspore culture and anther culture) whilst wheat DHs are produced via interspecific hybridization (wheat × maize wide crossing) and androgenic methods (wheat anther culture).

1.1 Androgenesis

Both isolated microspore culture and anther culture have been used to produce large numbers of barley DH lines for breeding and research programs in Australia. The Australian protocols have been used over a wide range of spring barley genotypes with good success. Although the production of albino plants has been cited as a problem for barley [12], it is generally not a limiting factor in Australian material. In wheat however, DH production across a broad range of genotypes using microspore-based methods is still not routine; wheat genotypes can vary in their androgenic response and the excessive production of albino plants can also be a problem. These issues have been experienced with Australian material [13, 14] and are also experienced in other programs [2, 15].

As with all androgenic methods, there are a number of important steps in the process which impact on the production of green regenerant plants, including: donor plant growth, spike selection, microspore stage, pretreatment, induction and regeneration medium and culture conditions. The use of controlled environment rooms for donor plant growth is an important feature of the Australian programs and allows for consistent year-round DH production, although donor plants can also be grown successfully in glasshouses and the field in some environments. Spikes are generally selected for anther or microspore culture when microspores are at the mid-uninucleate stage for microspore culture (barley) and mid-late to late uninucleate for anther culture (barley and wheat).

The pretreatment phase is the key step in reprogramming microspores from a pathway of gametogenesis to embryogenesis and plant development. Pretreatment may involve treatment of tillers, excised spikes, or anthers, and a variety of physical and chemical stresses are employed to induce this reprogramming event, sometimes in combination for maximum effect. Common stresses include heat shock, cold shock, and carbohydrate starvation. Following pretreatment, isolated microspores or anthers are transferred to an induction medium for embryo development. Embryos are subsequently transferred to a regeneration medium for plant development and sometimes to a rooting medium for root development. When root development is sufficient, plants are transferred to soil, usually seedling trays in a protected environment, followed by transfer to a glasshouse or the field.

A vital step in producing DH plants is to restore the chromosome number from haploid to fertile diploid status. In plants generated from anther and microspore culture, chromosome doubling may occur spontaneously and no chemical treatments are required. In barley the frequency of spontaneous doubling is relatively high and many programs report doubling rates between 66 and 90 % for both anther and microspore culture [16, 17]. Because of the consistently high doubling rates in barley, many DH programs do not apply colchicine, preferring instead to plant out additional regenerant plants to generate the required number of DHs. Morphological observations of mature plants are used to discard haploids and tetraploids. In other programs, ploidy is measured using rapid flow cytometry techniques, and haploid plants are treated with colchicine to double chromosome number [18].

Spontaneous doubling rates in wheat tend to be both lower and more variable than in barley. Variation between 24 and 80 % has been observed in Australian genotypes [19] and frequencies of less than 40 % have been reported is other studies [20, 21]. There have been two approaches to this issue. The first has been to treat cultures with colchicine in vitro, to induce early genome doubling and avoid some of the problems of traditional root-dipping colchicine treatments such as ploidy chimeras and variable seed set. Significantly higher rates of chromosome doubling have been obtained in wheat plants derived from both anther and microspore culture using in vitro colchicine treatments [22–24]. The second approach is the same as barley where wheat regenerant plants are measured for ploidy and then colchicine is applied to haploid plants using traditional root-dipping techniques. High throughput flow cytometry techniques using Partec Cyflow Space machinery, in connection with an autosampler for 96-well microtiter plates, enable up to 1,000 wheat plants to be screened per day [25].

1.2 Interspecific Crossing and Paternal Chromosome Elimination

An alternative and very effective method of producing wheat doubled haploids involves chromosome elimination following interspecific pollination of wheat by maize. The maize chromosomes are eliminated during the early divisions of the young embryo resulting in a haploid wheat embryo. Rescue of the embryos to artificial culture medium is required since endosperm usually fails to develop [26]. Wheat × maize crosses have been studied since Zenkteler and Nitsche [27] discovered that embryos were formed after pollination of wheat with maize. Pollination of wheat with maize, followed by an auxin treatment has become a well-established method of obtaining haploid wheat plants [28, 29].

Haploid plants derived from interspecific hybridization always require the application of anti-mitotic chemicals, such as colchicine, to double chromosome number and restore fertility. Although colchicine can be applied to embryos [30], it is most commonly applied to haploid plants. The method described by Jensen [31] for barley haploids at the 5-leaf stage, involving a 5 h root wash treatment with 0.05 % colchicine and 1.5 % DMSO, is widely used for cereals. Alternatively, as described below, the treatment time can be extended and the colchicine concentration reduced. Although doubling rates of 80–90 % can be achieved with these colchicine treatments, doubling success is very dependent on the environment in which the plants are grown and there are often ploidy chimeras resulting in variable seed set [17, 32]. Colchicine is also highly toxic and can result in high plant mortality rates and requires careful handling by operators.

The following sections describe in detail the protocols used at the South Australian Research and Development Institute (SARDI) and the Department of Agriculture and Food WA (DAFWA) to produce barley and wheat doubled haploids for commercial breeding programs and mapping projects.

2 Materials

There are four protocols described in this chapter: (1) isolated microspore culture (barley); (2) anther culture (barley); (3) anther culture (wheat); and (4) interspecific hybridization (wheat). The different plant tissue culture media used for each protocol

Table 1
Medium required for barley isolated microspore culture

Components	KFWC induction medium (liquid) (mg/L)	KFWC induction medium (solid) (mg/L)	Regeneration medium (MS basal) (mg/L)
Macro and micro salts			
KNO_3	1,900.0	1,900.0	1,900.0
NH_4NO_3	165.0	165.0	1,650.0
$CaCl_2 \cdot 2H_2O$	440.00	440.00	440.0
$MgSO_4 \cdot 7H_2O$	370.0	370.0	370.0
KH_2PO_4	170.0	170.0	170.0
$MnSO_4 \cdot 4H_2O$	22.3	22.3	22.3
$ZnSO_4 \cdot 7H_20$	8.6	8.6	8.6
H_3BO_3	6.2	6.2	6.2
KI	0.83	0.83	0.83
$Na_2Mo_4 \cdot 2H_2O$	0.25	0.25	0.25
$CoCl_2 \cdot 6H_2O$	–	–	0.025
$CuSO_4 \cdot 5H_2O$	0.025	0.025	0.025
Iron source			
$FeNa_2EDTA$	40.0	40.0	–
Na_2EDTA	–	–	33.5
$FeCl_3 \cdot 6H2O$	–	–	27.0
Other components			
Glutamine	600.0	600.0	–
Casein hydrolysate	300.0	300.0	–
Myo-inositol	200.0	200.0	100.0
Citric acid	10.0	10.0	
Pyruvic acid	10.0	10.0	
Glycine			2.0
Nicotinic acid	2.0	2.0	0.5
Pantothenic acid	2.0	2.0	
Pyridoxine HCl	2.0	2.0	0.5
Thiamine HCl	0.4	0.4	0.1
Biotin	0.02	0.02	–
IAA	1.0	1.0	–
BA	1.0	1.0	–
Cefotaxime	200.0	–	–
Maltose	60.0 or 90.0 g/L	60.0 g/L	–
Sucrose	–	–	20.0 g/L
Agar (Phytotechnology Laboratories A111)	–	–	7.0 g/L
Phytagel (Sigma P8169)	–	3.0 g/L	–
pH	5.8	5.8	5.9

are summarized (*see* Subheadings 2.1–2.4). Other equipment and materials required for these protocols is additionally outlined (*see* Subheading 2.5).

2.1 Barley Isolated Microspore Culture (IMC)

1. *KFWC Induction medium (liquid)*: Mix listed ingredients (*see* Table 1) by combining stock solutions and dry components (*see* **Note 1**) and make up to the desired volume using

distilled water. Mix, pH adjust, then filter-sterilize using a Steritop 0.22 µm filter (EMD Millipore Corporation, Billerica, MA, USA). Store in sterile Schott bottles at 4 °C until required. This medium is based on "medium C" of Kuhlmann and Foroughi-Wehr [33] and is prepared as a liquid for barley IMC induction and in solid form for plantlet regeneration (*see* Subheading 2.1 **item 2**).

2. *KWFC Induction medium* (*solid*): Prepare 500 mL of double strength (0.6 %) Phytagel (Sigma Chemical Company, St. Louis, MO, USA) using distilled water and autoclave at 121 °C (*see* **Note 2**). Allow mixture to cool to ~50 °C. Add this mixture 1:1 to double strength filter-sterilized liquid KFWC induction medium (*see* Subheading 2.1 **item 1**). Pour into 100 mm square petri dishes.
3. *Murashige and Skoog's* (*MS*) *basal medium*: Combine components (*see* Table 1) with distilled water to make up the desired final volume, mix, pH-adjust, and then autoclave at 121 °C (*see* **Note 2**). Whilst medium is still hot, decant into sterilized Magenta vessels (Sigma). Dry MS basal medium components can be sourced pre-mixed. This medium is used for root induction.

2.2 Barley Anther Culture

1. *Mannitol pretreatment* (*PT*) *medium*: This medium is supplemented with 1 M mannitol and used for anther pretreatment [33, 34]. The addition of 5.9 g/L $CaCl_2{\cdot}2H_2O$ improves embryo and green plant regeneration [35, 36]. Combine components (*see* Table 2) with distilled water to make up the desired final volume, autoclave at 121 °C (*see* **Note 2**) then pour into 55 × 14 mm petri dishes. This medium is not pH adjusted.
2. *Induction medium*: This medium is based on Hunter's FHG medium [37] with minor modifications: potassium iodide is included (0.83 mg/L), glutamine levels are slightly elevated (750 mg/L), maltose levels are increased (100 g/L) and BA and IAA are included (1.0 mg/L) (*see* Table 2.). Combine stock solutions and dry components (*see* **Note 1**) with distilled water to make up half the final volume. Mix, pH adjust then filter-sterilize through a Steritop 0.22 µm filter. The other half of this medium is prepared separately, by combining agar with distilled water. Autoclave this solution at 121 °C (*see* **Note 2**) and then allow to cool to ~50 °C. Mix with the filter-sterilized portion and pour into 55 × 14 mm petri dishes.
3. *Regeneration medium* (*MS basal*): Combine components (*see* Table 2) with distilled water to make up the desired final volume, mix, pH-adjust, and then autoclave at 121 °C (*see* **Note 2**). Whilst medium is still hot, decant into 100 × 60 mm gamma-sterilized tubs (FPA Australia, Sydney, NSW, Australia). Dry MS basal medium ingredients can be sourced pre-mixed. This medium is used for plantlet regeneration/root induction.

Table 2
Medium required for barley anther culture

Components	Mannitol PT medium (mg/L)	Induction medium (mg/L)	Regeneration medium (MS basal) (mg/L)
Macro and micro salts			
KNO_3	–	1,900.0	1,900.0
NH_4NO_3	–	165.0	1,650.0
$CaCl_2 \cdot 2H_2O$	5,900.0	440.0	440.0
$MgSO_4 \cdot 7H_2O$	–	370.0	370.0
KH_2PO_4	–	170.0	170.0
$MnSO_4 \cdot 4H_2O$	–	22.3	22.3
$ZnSO_4 \cdot 7H_2O$	–	8.6	8.6
H_3BO_3	–	6.2	6.2
KI	–	0.83	0.83
$Na_2MoO_4 \cdot 2H_2O$	–	0.25	0.25
$CoCl_2 \cdot 6H_2O$	–	0.025	0.025
$CuSO_4 \cdot 5H_2O$	–	0.025	0.025
Iron source			
$FeNa_2EDTA$	–	40.0	–
Na_2EDTA	–	–	37.26
$FeSO_4 \cdot 7H_2O$	–	–	27.80
Other components			
Glutamine	–	750.0	–
Myo-inositol	–	100.0	100.0
Glycine	–	–	2.0
Nicotinic acid	–	–	0.5
Pyridoxine HCl	–	–	0.5
Thiamine HCl	–	0.4	0.1
IAA	–	1.0	–
BA	–	1.0	–
Maltose	–	100.0 g/L	–
Sucrose	–	–	20.0 g/L
Mannitol	182.0 g/L	–	–
Agar (Sigma A7921)	20.0 g/L	10.0 g/L	–
Phytagel (Sigma P8169)	–	–	3.0 g/L
pH	Not adjusted	5.8	5.8

2.3 Wheat Anther Culture

1. *Mannitol Pretreatment (PT) medium*: *see* Table 3 and prepare according to Subheading 2.2 **item 1** with the exception that macronutrients are also included [19, 38].
2. *Liquid induction medium (LIM)*: Mix listed ingredients (*see* Table 3) by combining stock solutions and dry components (*see* **Note 1**) and make up to the desired volume using distilled water. Mix, pH adjust, and then filter-sterilize using a Steritop 0.22 μm filter. Store in sterile Schott bottles at 4 °C until required. The development of this liquid induction medium is described by Broughton [14].

Table 3
Medium required for wheat anther culture

Components	Mannitol PT medium (mg/L)	Liquid induction medium (LIM) (mg/L)	Regeneration medium (mg/L)
Macro and micro salts			
KNO_3	1,400.0	1,400.0	1,400.0
NH_4NO_3	300.0	300.0	300.0
$CaCl_2{\cdot}2H_2O$	6,340.0	440.0	440.0
$MgSO_4{\cdot}7H_2O$	370.0	370.0	370.0
KH_2PO_4	170.0	170.0	170.0
$MnSO_4{\cdot}4H_2O$	–	22.3	22.3
$ZnSO_4{\cdot}7H_20$	–	8.6	8.6
H_3BO_3	–	6.2	6.2
KI	–	0.83	0.83
$Na_2Mo_4{\cdot}2H_2O$	–	0.25	0.25
$CoCl_2{\cdot}6H_2O$	–	0.025	0.025
$CuSO_4{\cdot}5H_2O$	–	0.025	2.5
Iron source			
$FeNa_2EDTA$	–	40.0	40.0
Other components			
Glutamine	–	750.0	750.0
N-Z-Amine A	–	300.0	300.0
Myo-inositol	–	100.0	100.0
Proline	–	100.0	100.0
Citric acid	–	10.0	10.0
Pyruvic acid	–	10.0	10.0
Glycine	–	2.0	2.0
Nicotinic acid	–	2.0	2.0
Pantothenic acid	–	2.0	2.0
Pyridoxine HCl	–	2.0	2.0
Thiamine HCl	–	0.4	0.4
Biotin	–	0.02	0.02
Gum arabic (Sigma G9752)	–	10.0	–
Larcoll (Aldrich 85,1361)	–	10.0	–
2,4-D	–	2.0	–
Kinetin	–	0.5	–
Cefotaxime	–	100.0	–
Maltose	–	100.0 g/L	30.0 g/L
Mannitol	182.0 g/L	–	–
Agar (Sigma A7921)	15.0 g/L	–	–
Phytagel (Sigma P8169)	–	–	3.0 g/L
pH	Not adjusted	5.8	5.8

3. *Regeneration medium*: *see* Table 3 and prepare according to Subheading 2.2 **item 2** with the exception that this medium is poured into 90 × 20 mm petri dishes. The development of this solid regeneration medium is described by Broughton [14].

Table 4
Medium required for interspecific hybridization (wheat × maize wide crossing)

Components	Embryo rescue medium (mg/L)
Macro and micro salts	
KNO_3	2,500.0
$CaCl_2 \cdot 2H_2O$	150.0
$MgSO_4 \cdot 7H_2O$	250.0
$NaH_2PO_4 \cdot 2H_2O$	170.0
$(NH_4)_2SO_4$	134.0
$MnSO_4 \cdot 4H_2O$	13.7
$ZnSO_4.7H_20$	2.0
H_3BO_3	3.0
KI	0.75
$Na_2Mo_4 \cdot 2H_2O$	0.25
$CoCl_2 \cdot 6H_2O$	0.025
$CuSO_4 \cdot 5H_2O$	0.025
Iron source	
Na_2EDTA	37.25
$FeSO_4 \cdot 7H_2O$	27.85
Other components	
Myo-inositol	100.0
Nicotinic acid	1.0
Pyridoxine HCl	1.0
Thiamine HCl	10.0
Sucrose	20.0 g/L
Agar (Sigma A7049)	10.0 g/L
pH	6.0

2.4 Wheat Interspecific Hybridization

1. *Wheat embryo rescue medium*: is based on Gamborg's B5 basal medium [39] and supplemented with sucrose. Mix listed components (*see* Table 4) in an appropriately sized beaker and make up to 500 mL with distilled water. Mix and pH adjust. In a separate beaker dissolve agar in 500 mL of distilled water whilst mixing on a hotplate. Combine both solutions thoroughly and aliquot into Biju or McCartney bottles before autoclaving at 121 °C (*see* **Note 2**). After autoclaving, McCartney bottles are placed on a 45° angle to set medium on a slope. Dry B5 basal medium components can be sourced pre-mixed.

2.5 Additional Materials

1. *Cell counter*: Hemocytometer.
2. *Chemicals*: carmine for acetocarmine stain, cefotaxime, colchicine, *n*-butanol and MES buffer (2-[*N*-morpholino]ethanesulfonic acid).
3. *Controlled environment rooms*: constructed by Phoenix Biosytems (Edwardstown, SA, Australia) with light provided

by Osram 400 watt Vialox Nav-T (SON-T) lamps (OSRAM Australia, Pennant Hills NSW, Australia).

4. *Crossing bags*: Glassine paper bags.
5. *Fertilizers*: Osmocote Exact, Osmocote Plus, Peters Professional (Scotts Australia Pty. Ltd., Bella Vista, NSW, Australia), Phostrogen (Bayer Cropscience, Cambridge, UK).
6. *Filter units*: Steritop-GP Filter Unit 0.22 μm filter.
7. *General laboratory equipment*: fine forceps, scissors, scalpels, glassware including Schott bottles, balance, hot-plate/stirrer, refrigerator (4 °C), freezer (−20 °C), fume cabinet, laminar flow cabinet, growth cabinet, Bunsen burner, water-bath, pH meter, dissecting microscope, light microscope, inverted microscope (for isolated microspore protocol) and access to distilled water.
8. *Lamps*: Fluorescent lamps for growth cabinets (cool white and Grolux).
9. *Microblender chamber*: Model 8580 (Eberbach Corporation, Ann Arbor, MI, USA).
10. *Petri dishes, vessels, and sealers*: 55 × 14 mm petri dishes, 90 × 14 mm petri dishes, and 100 mm square petri dishes, 90 × 20 mm petri dishes, Magenta vessels (Sigma), 100 × 60 mm gamma-sterilized tubs (FPA Australia), Biju or McCartney bottles, Parafilm™.
11. *Seed trays*: Kwik pots (20 cell) (Garden City Plastics, Monbulk, VIC, Australia).

3 Methods

3.1 SARDI Barley Isolated Microspore Culture (IMC)

3.1.1 Donor Plant Growth Conditions

1. Maintain controlled environment rooms for donor plant growth at 17/14 °C (day/night) with a 14 h photoperiod. Lighting is provided by 400 W Osram sodium vapor lamps (450–500 μmol/m^2/s at canopy level).
2. Sow seeds directly into pots containing 90 % Coco peat, 10 % sharp nursery cuttings sand supplemented with 5.4 g/L of "Osmocote Plus" slow release fertilizer (*see* **Note 3**).
3. After 4 weeks, fertilize donor plants weekly with liquid Phostrogen (0.5 g/L water) (Bayer Cropscience). Phostrogen contains 14:4.4:22.5 (N:P:K elemental analysis) plus trace elements.

3.1.2 Spike Collection, Dissection and Pretreatment (Cold)

1. For IMC, spikes are selected containing microspores at the early to mid-uninucleate stage of development (*see* **Note 4**). Spikes ready to harvest are identified when the base of the flag leaf has emerged about 4–7 cm beyond the preceding leaf sheath, approximately 9–11 weeks after sowing.

2. Cut tillers several nodes below the spike and place in a beaker containing distilled water. These can be stored at 4 °C until the spikes are dissected (0–2 days later).
3. Prior to spike dissection, remove outer leaves (blade and sheath) from the tiller and swab inner leaves with 70 % ethanol. Leave tillers to dry for 2–3 min in a laminar flow cabinet.
4. Dissect spikes aseptically from the tillers, trim awns and place in 90 × 14 mm petri dishes (4–6 spikes per dish). Add a small piece of sterile dish-washing sponge saturated with sterile water to each petri dish to maintain humidity in the dishes. Seal petri dishes with Parafilm™.
5. Spikes are stored in petri dishes in the dark at 4 °C for 2–4 weeks before microspore isolation.

3.1.3 Culture Initiation: Microspore Extraction, Purification and Embryo Induction

1. Following cold pretreatment of spikes, remove anthers from 7 to 10 spikes under aseptic conditions and placed in a petri dish containing 5–10 mL of KFWC Induction medium supplemented with 60 g/L of maltose.
2. Place anthers and medium in a microblender chamber (Model 8580, Eberbach, Mich., USA). Before blending, make the volume up to 20 mL with additional KFWC medium. Blend at high speed for 20 s.
3. Filter the blended mixture through a 100 μm sieve and centrifuge at 100 × *g* for 10 min.
4. Remove the supernatant and resuspend microspores in a solution containing 580 mM maltose, 5 mM MES buffer (2-[*N*-morpholino]ethanesulfonic acid) with a pH of 5.8. Make up to a final volume of approximately 11 mL and pour into a sterile 10 mL volumetric flask.
5. Using a sterile Pasteur pipette, gently layer 1 mL of 300 mM mannitol and 5 mM MES (pH 5.8) onto the maltose solution containing the microspores. Cap the volumetric flask with sterile aluminum foil and centrifuge at 85 × *g* for 8 min. Centrifugation settles the viable microspores between the maltose and mannitol layers.
6. Remove the settled microspores with a sterile Pasteur pipette. The total volume to be removed is approximately 500 μL. Make the solution containing microspores up to 5 mL with KFWC Induction medium supplemented with 90 g/L of maltose.
7. Estimate the total number of microspores by counting cells contained within a 1 μL aliquot. This can be done using a hemocytometer and observing under a microscope at 400× magnification.
8. Centrifuge the microspore solution at 100 × *g* for 10 min, remove the supernatant, and adjust microspore concentration

to 1×10^5 microspores per mL using KFWC Induction medium supplemented with 90 g/L maltose.

9. Plate microspores out aseptically in either 90 × 14 mm diameter petri dishes (6 mL per dish) or 55 × 14 mm petri dishes (2 mL per dish). Supplement the KFWC Induction medium with 200 μg/mL of cefotaxime (antibiotic). Seal dishes with Parafilm™ and incubate in the dark at 22–25 °C for 21–28 days. Cefotaxime is effective in controlling bacterial contamination and has no adverse effects on barley isolated microspore cultures.

3.1.4 Plantlet Regeneration, Root Induction and Transfer to Soil

1. After 4–5 weeks transfer microspore derived colonies to plates containing KFWC Induction medium (60 g/L maltose) solidified with 0.3 % Phytagel (Sigma). Optimal density is 12–25 colonies per cm^2. Seal dishes with Parafilm™.
2. Incubate cultures in a growth cabinet at 22–25 °C under a 16 h photoperiod with a lighting intensity of 50 μmol/m^2/s using a combination of cool white and Grolux fluorescent lamps. Shoot growth is generally visible within 2–4 weeks.
3. Transfer green plantlets to MS basal medium in Magenta vessels to promote root growth before transfer to soil. Allow 3–4 weeks to stimulate root growth.
4. Well-rooted plantlets are transferred to 50 mm seedling tubes and grown in controlled environment rooms (*see* Subheading 3.1.1) for 3–4 weeks.
5. Place tubes on a bed of potting mix soil. This soil should be approximately 80 mm deep and contained within plastic boxes containing drainage holes. This prevents soil from waterlogging. Twenty plants can be grown per box, each of which can produce several spikes. This is an efficient method of growing a large number of plants in a relatively small area (*see* Fig. 1). At maturity, fertile doubled haploid lines are harvested (*see* **Note 5**).

3.2 DAFWA Barley Anther Culture

3.2.1 Donor Plant Growth Conditions

1. Maintain controlled environment rooms for donor plant growth at 18/13 °C (day/night) with a 12-h photoperiod. Sodium vapor lamps (Osram 400 W) provide approximately 450–500 μmol/m^2/s of photosynthetically active radiation (PAR) at canopy level.
2. Sow seeds directly into 5.2 L pots (two plants per pot) containing a commercial potting mix (*see* **Note 6**). Fertilize weekly with a commercial soluble fertilizer such as Peters Professional (Scotts Australia Pty. Ltd) which contains 27:5.5:9 (N:P:K elemental analysis).

Fig. 1 Barley isolated microspore derived plants growing in separate seedling tubes within plastic boxes containing a layer of potting soil to promote root growth and higher seed yield

3.2.2 Spike Collection and Dissection

1. Follow as per SARDI IMC method (*see* Subheading 3.1.2) with the exception that for anther culture, spikes are selected when microspores are at the mid-late to late uninucleate stage (*see* **Note 4**) and mannitol pretreatment of anthers replaces cold pretreatment of spikes.
2. Mannitol pretreatment of anthers should ideally occur on the same day as spike collection.

3.2.3 Anther Pretreatment (Mannitol)

1. Excised spikes are placed into 90 × 14 mm petri dishes, approximately 4–5 spikes per dish, and stored at 4 °C until anther removal. This should be performed on the same day as spike dissection (*see* Subheading 3.2.2).
2. Remove anthers from each floret and place on solid mannitol PT medium in 55 × 14 mm petri dishes (one spike per dish).
3. Dissect between 50 and 75 anthers per spike. This number will depend on the length of the spike and quality of the anthers.
4. Seal dishes with Parafilm™ and incubate in the dark at 25 °C for 4–5 days.

3.2.4 Embryo Induction

1. Following mannitol pretreatment anthers are transferred to 55 × 14 mm petri dishes containing solid induction medium, sealed with Parafilm™ and incubated in the dark at 25 °C for 5–6 weeks.

3.2.5 Plantlet Regeneration, Root Induction and Transfer to Soil

1. Transfer anther culture derived embryos and calli to tubs containing MS Basal medium (*see* Table 2) and incubate in a growth cabinet under a 12 h photoperiod with low light (50 μmol PAR/m^2/s). These conditions are provided by a combination of cool white and Grolux fluorescent lamps. Maintain at a temperature of 25 °C for 4–5 weeks. During this time, shoots and roots develop and plantlets are usually ready for transfer to soil in approximately 4 weeks.
2. Rooting plantlets are transferred into 20-cell seedling trays (Kwik pots, Garden City Plastics) containing soil. Clear plastic lids are placed on these trays for the first week after transplanting to facilitate hardening-off.
3. Fertilize plants with a slow release fertilizer such as Osmocote Exact (Scotts Australia Pty. Ltd.), and maintain plants in controlled environment rooms (18/13 °C) for at least 3–4 weeks before transfer to either the field or glasshouse.
4. Once sufficient biomass has been generated, plants can be transferred to either a glasshouse and grown to maturity in 20-cell seedling trays, or field conditions (*see* **Note 5**). If transplanting to the field, this should preferentially occur during the typical growing season for local cereal varieties. In Australia this is during the winter months (May-July). The advantage of field grown plants is that significantly greater seed volumes are produced and the DH lines do not require a subsequent bulk-up phase.

3.3 DAFWA Wheat Anther Culture

3.3.1 Donor Plant Growth Conditions

1. Wheat donor plants are grown in controlled environment rooms under similar conditions as barley (18/13 °C, 12 h photoperiod) (*see* Subheading 3.2.1), with the exception that one plant is sown per pot instead of two.

3.3.2 Spike Collection and Dissection

1. Harvest wheat spikes for anther culture when the leaf sheath surrounding the spike is beginning to open (~1–2 mm). Microspores should be at the mid-late to late uninucleate stage of development (*see* **Note 4**).
2. After harvest, spikes are either sterilized and dissected immediately or stored in a beaker of water for between 1 and 4 days at 4 °C. This time can be extended to 5–7 days, depending on workload.
3. Sterilize spikes as per Kasha et al. [40]: trim awns and immerse spikes for 15 min in commercial bleach diluted to 10.5 g/L sodium hypochlorite.
4. Rinse spikes in sterile distilled water for 5 min twice.
5. Allow spikes to dry in an open petri dish placed in a laminar flow cabinet for 10–30 min prior to dissection.

3.3.3 Pretreatment (Mannitol)

1. Dissect thirty anthers from either side of each spike and place on solid mannitol pretreatment medium in 55 × 14 mm petri dishes. Place anthers from each spike in a separate dish.
2. At the same time as the anthers are dissected, remove ten ovaries from each spike and place in a separate 55 × 14 mm petri dish containing 4 mL liquid induction medium (LIM). Thus for each spike there are two dishes, one containing anthers and the other containing LIM + ovaries.
3. Seal all petri dishes with Parafilm™ and incubate in the dark at 25 °C for 5 days.

3.3.4 Pretreatment (n-Butanol)

1. Prepare a fresh set of 55 × 14 mm petri dishes with 4 mL LIM supplemented with 8 μL *n*-butanol (0.2 % v/v).
2. Remove anthers from each dish of solid mannitol medium and place onto a dish containing LIM + *n*-butanol. Leave dishes unsealed and incubate in the dark at ~25 °C for 5 h. Treatment of anthers with *n*-butanol has been shown to significantly increase embryo and green plantlet production in wheat anther culture [19, 41].
3. Remove the LIM + *n*-butanol from each dish by pipette, leaving the anthers behind. Take the dish of LIM + ovaries from the same spike and tip the LIM (now "ovary-conditioned") and ovaries into the dish containing the anthers. If any ovaries are left behind, use forceps to gently transfer them.
4. Seal dishes with Parafilm™. Gently swirl the dishes so that all anthers are floating and not stuck to the sides or bottom of the dish.

3.3.5 Embryo Induction, Regeneration and Root Induction

1. Following *n* butanol treatment, anthers and ovaries are co- incubated in the dark at 25 °C for 4–6 weeks.
2. Embryos will be visible after 4–6 weeks.
3. Transfer wheat anther culture derived embryos to solid regeneration medium in 90 × 20 mm petri dishes (*see* Fig. 2). Usually embryos from each 55 × 14 mm dish are tipped onto one or two dishes, depending on embryo density.
4. Seal petri dishes with Parafilm™ and incubate in the dark at 25 °C for 7 days, then transfer to a growth cabinet with a 12 h photoperiod. Growth conditions are the same as for barley anther culture (*see* Subheading 3.2.5). Incubate for 3–4 weeks to stimulate shoot and root growth.
5. After 4 weeks, or once plantlets have grown substantial root biomass, they are transplanted to 20-cell seedling trays ("Kwik pots") containing soil. Clear plastic lids are placed on these trays for the first week after transplanting to facilitate hardening-off. Maintain plants in controlled environment rooms (18/13 °C) for 3–4 weeks, then transfer to the glasshouse, grow to maturity and harvest DH lines (*see* **Note 7**).

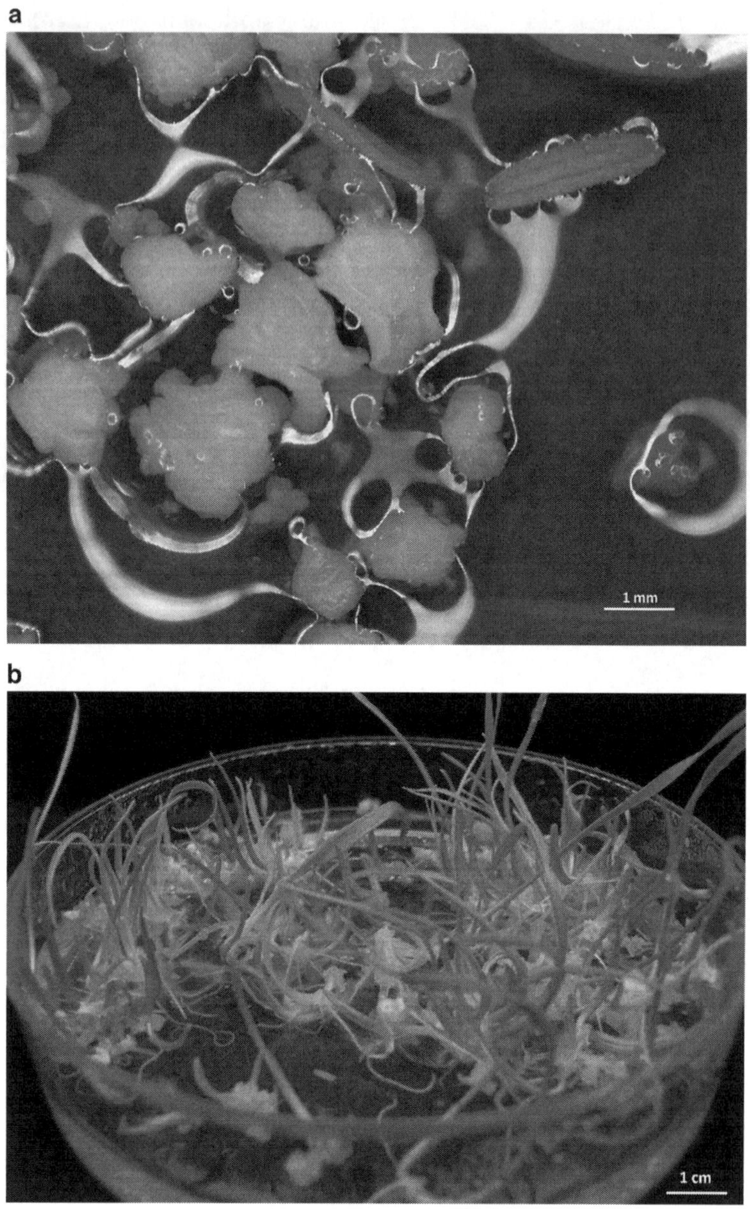

Fig. 2 (**a**) Wheat embryos after 4–6 weeks' incubation in LIM plus ovaries. (**b**) Green and albino plants develop after embryos are transferred to solid regeneration medium

3.4 SARDI Interspecific Hybridization (Wheat × Maize Wide Crossing)

3.4.1 Donor Plant Growth Conditions

1. Grow donor wheat plants in controlled environment rooms at 17/14 °C day/night with a 14 h photoperiod under 400 watt sodium vapor lamps. Plants should be grown in appropriate potting mix soil; here we use SARDI "DH donor potting mix" as previously described (*see* Subheading 3.1.1).
2. Grow maize pollinators under glasshouse conditions keeping the temperature at approximately 20/14 °C (day/night) with natural day length. Lighting is supplemented during the winter

months with metal halide lamps. Many varieties of maize will provide successful results but for consistency we use the same variety of sweet corn, Kelvedon Glory F1.

3.4.2 Emasculation and Pollination

1. Select wheat spikes 1–2 days before anthesis and emasculate. These spikes can also be identified by the coloration of the anthers as they are almost completely bright yellow and there is no sign of residual green coloration. Generally spikes ready for emasculation are present approximately 8 weeks post planting, however this is dependent upon the cultivars used. Selecting primary spikes is preferable.
2. Place glassine paper bag over emasculated spike and leave for 2–3 days.
3. Remove glassine paper bag and inspect for florets containing elongated and separated stigmatic hairs (feathery and fluffy). These florets are ready for pollination. If florets of this form are not identified then leave bagged for a further 2–4 days.
4. Collect pollen on a piece of folded paper and apply to emasculated florets using a small sable hair paintbrush.
5. Return glassine paper bags over the spikes immediately post pollination to prevent donor plant inter-crossing.
6. Spray a solution of 2,4-D (0.01 % w/v) onto pollinated spikes 24–48 h post pollination.
7. Remove glassine paper bags 48 h after spraying.

3.4.3 Spike Harvest and Embryo Rescue

1. Harvest spikes 17–20 days after the first pollination (this is cultivar dependent) by cutting the stem 10–20 cm below the base of the spike and placing in a beaker of distilled water. Change water and trim the base of the stem daily. Although it is preferential to excise embryos the same day as harvesting, spikes may be stored at 4 °C for up to 7 days.
2. Carefully remove immature seeds with tweezers and place in sterilized McCartney bottles.
3. Rinse seeds in 70 % ethanol for approximately 30 s then remove ethanol and replace with 10–15 mL of bleach (1 % active chlorine). Leave for 5 min with intermittent agitation. Remove bleach and wash twice with sterilized distilled water. Place seeds in a sterile petri dish under aseptic conditions.
4. Open seeds aseptically with sterile tweezers using a dissecting microscope. Transfer embryos to embryo rescue medium, immature scutella facing up, whilst maintaining an aseptic technique.
5. Incubate embryos at 4 °C in the dark for 2 days followed by 2 days in the dark at room temperature.
6. Transfer embryos to a growth cabinet (22 °C, 16 h photoperiod) for 2–3 weeks to stimulate growth. If no growth is

observed over a period of 2 weeks, they are returned to 4 °C for 2–3 days. They are subsequently transferred back to the growth cabinet.

3.4.4 Transplanting Haploid Plantlets from Agar to Soil

1. Transfer haploid plantlets to 50 mm seedling tubes (50 × 65 mm) containing soil saturated with fertilizer solution (Phostrogen, 0.5 g/L).
2. Place tubes containing plantlets in seedling trays (40 tubes/tray). These trays are transferred to a controlled environment room set at between 14 and 17 °C (*see* Subheading 3.4.1) for 3–4 weeks.

3.4.5 Chromosome Doubling (Colchicine Treatment)

1. In this method, all plants are haploid and must be treated with colchicine to double chromosome numbers and restore fertility. Select plants producing at least 4 tillers and wash soil from roots.
2. Remove the oldest leaves to expose young shoots and trim back so that shoots are 10–15 cm in length. Trim roots back to 1 cm in length.
3. Place trimmed plants in water to minimize dehydration stress. Transfer to a fume cabinet.
4. Using appropriate safety precautions, prepare a 0.4 % colchicine stock (this solution can be stored at 4 °C for up to 2 months). Dilute to a working colchicine solution of 0.026 % by combining 100 mL of colchicine stock solution, 15 mL DMSO and 1,385 mL of reverse osmosis water in a 2 L beaker.
5. Blot excess water from trimmed plants using paper towels.
6. Place trimmed plants (~120 at a time) in 150 mL of working colchicine solution. The best vessel for this is a narrow yet tall 500 mL beaker. Ensure roots are fully immersed. Maintain immersed plants at 22 °C in a fume cabinet with continuous aeration provided by an aeration pump (*see* Fig. 3). Leave for 16–18 h. This working solution can be used up to five times, topping it up with approximately 50 mL fresh solution each time and keeping at 4 °C when not in use.
7. Remove plants from colchicine solution and wash continuously under running tap water for more than 1 h.

3.4.6 Transfer to Soil and Recovery of Doubled Haploid Plants

1. Transfer colchicine-treated plants to 50 mm seedling tubes containing soil. Place in controlled environment rooms (*see* Subheading 3.4.1) for approximately 1 week. This is the typical time frame for plants to recover from colchicine stress as identified by fresh shoot growth.
2. Transfer plants to a greenhouse to stimulate vigorous growth.

Fig. 3 Aeration during wheat colchicine treatment

3. Once plants are observed to be growing vigorously transfer to tubs half filled with soil (as for barley DHs, *see* Fig. 1). This provides additional soil for root growth. Leave plants in these tubs till maturity and harvest DH lines (*see* **Note 8**).

4 Notes

1. Stocks are prepared for most media components and stored refrigerated at 4 °C (macro-salt, micro-salt and iron stocks) or frozen at −20 °C (vitamins, growth regulators and organic components). Stocks are replaced every 3–6 months. Other components such as glutamine, inositol, N-Z-Amine A (replacing casein hydrolysate), cefotaxime, and maltose are added at the time of media preparation. For the wheat anther culture induction medium, a stock of 2,4-D is prepared by dissolving 8 mg/mL 2,4-D in absolute ethanol (no water added). It is stored in the refrigerator and 0.25 mL stock is added per L of medium. Distilled water is used for all stock and medium preparation. Dry media components are sourced pre-mixed for

Murashige and Skoog's (MS) basal medium and Gamborg's B5 basal medium. Liquid media are stored refrigerated in the dark and used within 2–3 weeks whilst solid media are stored at room temperature in the dark and used within 4 weeks. Petri dishes are either resealed in bags or wrapped in Clingfilm during this time.

2. Autoclave times vary depending on the volume of medium; volumes of up to 500 mL are autoclaved for 20 min whilst volumes of 1 L are autoclaved for 30 min. Small volumes in McCartney bottles (~10 mL) are autoclaved for 15 min. If autoclaved portions are mixed with filter-sterilized portions, the autoclaved portion is first cooled to ~50 °C by placing bottles in a water-bath set between 50 and 55 °C for 30–60 min. The filter-sterilized portions can also be pre-warmed to 50 °C by placing bottles in the same water bath for a short period (15–30 min).
3. The SARDI "DH donor potting mix" is prepared with 450 L of Coco Peat and adding nursery cutting sand (sharp) to make a final volume of 500 L. To this is added 2.7 kg of Osmocote Plus slow release fertilizer (Scotts Australia Pty. Ltd.) and the pH is adjusted to between 6.0 and 6.5 using two parts agricultural lime to one part hydrated lime.
4. The developmental stage of barley and wheat microspores is checked in a small number of spikes at the beginning of the procedure for each cross and correlated with morphological indicators, which are used to select spikes for the remainder of spike harvest. Anthers from a central floret are squashed in a drop of 2 % w/v acetocarmine, covered with a cover-slip, and observed under 400× magnification. For anther culture, mid-late and late uninucleate microspores are preferred but good responses can also be obtained from slightly older spikes where mitosis is visible in some microspores.
5. In barley an average of 60–70 % of plantlets transferred to soil spontaneously double and are fully fertile, so chromosome doubling treatments are not required and tests for ploidy are not necessary. In crosses with a high regenerative capacity, this figure is often higher than 70 %, possibly because we have more plantlets to choose from and, in choosing the largest most vigorous plantlets, we are biasing selection in favor of doubled haploid plants and avoiding smaller haploid plants. Tetraploids are observed among the fertile spontaneously doubled plants but these occur at frequencies which can be ignored for the purposes of plant breeding because tetraploid barley plants generally perform poorly in the field and are eliminated early in the selection process. Under Australian growing conditions, the frequency of spontaneous chromosome doubling in barley is similar in plants derived from isolated microspore culture and anther culture.

6. The potting mix used at DAFWA, "Uni Field Station Mix", is prepared by Richgro Garden products (Jandakot, Perth, Australia). It contains equal parts pine bark compost (fine), coco peat (standard) and river sand (quartz). The following nutrients are included at 8.3 kg per cubic meter of mix: ammonium nitrate limestone (extra fine), dolomite (Watheroo Coarse), and Mineral Mix #4, while Superphosphate ("Better Grow") is included at 13.3 kg per cubic meter. The pH is adjusted to ~5.8–6.0.
7. In wheat, the rate of spontaneous doubling tends to be lower and more variable than barley and, as a result, some programs incorporate an in vitro colchicine treatment to induce early genome doubling. At DAFWA we have observed doubling frequencies between 24 and 80 % in anther culture derived wheat plants and the addition of colchicine to the induction medium, immediately after *n*-butanol treatment, has proven beneficial in increasing doubling frequencies in some crosses, but further evaluation is required. To date the protocol has involved adding either 50 mg/L colchicine for 5–7 days, or 100–200 mg/L for 2–4 days
8. On average, 60–70 % of colchicine-treated plants double successfully with this method; however, many spikes are not fully fertile and usually only 5–20 seeds per spike are obtained.

References

1. Khush GS, Virmani SS (1996) Haploids in plant breeding. In: Jain SR, Sopory SK, Veilleux RE (eds) In vitro haploid production in higher plants, vol 1, Fundamental aspects and methods. Kluwer Academic Publishers, Dordrecht, pp 11–33
2. Touraev A, Pfosser M, Heberle-Bors E (2001) The microspore: a haploid multipurpose cell. Adv Bot Res 35:53–109
3. Ascher-Ellis JS, Graham RD, Hollamby GJ et al (2001) In: Reynolds MP, Ortiz-Monasterio JI, McNab A (eds) Application of physiology in wheat breeding. CIMMYT, Mexico, DF, pp 219–240
4. Devaux P, Pickering R (2005) Haploids in the improvement of *Poaceae*. In: Palmer CE, Keller WA, Kasha KJ (eds) Biotechnology in agriculture and forestry—56 haploids in crop improvement II. Springer, Heidelberg, pp 215–242
5. Jauhar PP, Xu SS, Baenziger PS (2009) Haploidy in cultivated wheats: induction and utility in basic and applied research. Crop Sci 49:737–755
6. Collard BCY, Jahurfer MZZ, Brouer JB et al (2005) An introduction to markers, quantitative trait loci (QTL) mapping and marker-assisted selection for crop improvement: the basic concepts. Euphytica 142:169–196
7. Kammholz SJ, Campbell AW, Sutherland MW et al (2001) Establishment and characterisation of wheat genetic mapping populations. Aust J Agric Res 52:1079–1088
8. Cakir M, Gupta S, Li CD et al (2011) Genetic mapping and QTL analysis of disease resistance traits in the barley population Baudin × AC Metcalfe. Crop Pasture Sci 62:152–161
9. Howes NK, Woods SM, Townley-Smith TF (1998) Simulations and practical problems of applying marker-assisted selection and doubled haploids to wheat breeding programs. Euphytica 100:225–230
10. Bonnett DG, Rebetzke GJ, Spielmeyer W (2005) Strategies for efficient implementation of molecular markers in wheat breeding. Mol Breed 15:75–85
11. Kuchel H, Ye G, Fox R et al (2005) Genetic and economic analysis of a targeted marker-assisted wheat breeding strategy. Mol Breed 16:67–78
12. Devaux P, Kasha K (2009) Overview of barley doubled haploid production. In: Touraev A, Forster BP, Jain SM (eds) Advances in haploid

production in higher plants. Springer, Heidelberg, pp 47–63

13. Ding XL, Luckett DJ, Darvey NL (1995) A cost based index of anther culture response in diverse wheat-breeding germplasm. Aust J Exp Agric 35:395–401
14. Broughton S (2008) Ovary co-culture improves embryo and green plant production in anther culture of Australian spring wheat genotypes (*Triticum aestivum* L.). Plant Cell Tiss Org Cult 95:185–195
15. Tuvesson S, Ljungberg A, Johansson N et al (2000) Large-scale production of wheat and triticale doubled haploids through the use of a single-anther culture method. Plant Breed 119:455–459
16. Davies P (2003) Barley isolated microspore culture (IMC) method. In: Malszynnski M, Kasha KJ, Forster BP, Szarejko I (eds) Doubled haploid production in crop plants: a manual. Kluwer Academic Publishers, Dordrecht, pp 29–34
17. Weyen J (2009) Barley and wheat doubled haploids in breeding. In: Touraev A, Forster BP, Jain SM (eds) Advances in haploid production in higher plants. Springer, Heidelberg, pp 179–187
18. Jacquard C, Wojnarowiez G, Clement C (2003) Anther culture in barley. In: Malszynnski M, Kasha KJ, Forster BP, Szarejko I (eds) Doubled haploid production in crop plants: a manual. Kluwer Academic Publishers, Dordrecht, pp 21–27
19. Broughton S (2011) The application of *n*-butanol improves embryo and green plant production in anther culture of Australian wheat (*Triticum aestivum* L.) genotypes. Crop Pasture Sci 62:813–822
20. Ouyang J, Liang H, Jia S et al (1994) Studies on the chromosome doubling of wheat pollen plants. Plant Sci 98:209–214
21. Navarro-Alvarez W, Baenziger PS, Eskridge KM et al (1994) Addition of colchicine to wheat anther culture media to increase doubled haploid production. Plant Breed 112:192–198
22. Barnabás B, Pfahler PL, Kovacs G (1991) Direct effect of colchicine on the microspore embryogenesis to produce dihaploid plants in wheat (*Triticum aestivum* L.). Theor Appl Genet 81:675–678
23. Hansen NJP, Anderson SB (1998) In vitro chromosome doubling with colchicine during microspore culture in wheat (*Triticum aestivum* L). Euphytica 102:101–108
24. Soriano M, Cistué L, Vallés MP et al (2007) Effects of colchicine on anther and microspore culture of bread wheat (*Triticum aestivum* L.). Plant Cell Tiss Org Cult 91:225–234
25. Weyen J, Hoenemann C (personal communication)
26. Suenaga K, Nakajima K (1989) Efficient production of haploid wheat *T. aestivum* through crosses between Japanese wheat and maize *Zea mays*. Plant Cell Rep 8:263–266
27. Zenkteler M, Nitzsche W (1984) Wide hybridization experiments in cereals. Theor Appl Genet 68:311–315
28. Laurie DA, O'Donoughue S, Bennet MD (1990) Wheat x maize and other wide sexual hybrids: their potential for genetic manipulation and crop improvement. In: Gustavson JP (ed) Gene manipulation in plant improvement. Plenum, New York, pp 95–126
29. Suenaga K (1994) Doubled haploid system using the intergeneric crosses between wheat (*Triticum aestivum*) and maize (*Zea mays*). Bull Nat Inst Agrobiol Resour 9:83–139
30. Sood S, Dhawaan R, Singh K et al (2003) Development of novel chromosome doubling strategies for wheat × maize system of wheat haploid production. Plant Breed 122: 493–496
31. Jensen CJ (1974) Chromosome doubling techniques in haploids. In: Kasha KJ (ed) Haploids in higher plants. Advances and potential. Proceedings of the first international symposium. Guelph University Press, Guelph, pp 153–190
32. Kasha KJ (2005) Chromosome doubling and the recovery of doubled haploid plants. In: Palmer CE, Keller WA, Kasha KJ (eds) Biotechnology in agriculture and forestry—56 haploids in crop improvement II. Springer, Heidelberg, pp 123–147
33. Cistué L, Ramos A, Castillo AM et al (1994) Production of large numbers of doubled haploid plants from barley anthers pre-treated with high concentrations of mannitol. Plant Cell Rep 13:709–712
34. Broughton S, Priest P (1997) Barley doubled haploids in Western Australia—improved green plant regeneration from anthers pretreated with high concentrations of mannitol. Eighth Australian barley technical symposium, Gold Coast, Queensland, 7–12 Sep 1997
35. Hoekstra S, van Bergen S, van Brouwershaven IR et al (1997) Androgenesis in *Hordeum vulgare* L.: effects of mannitol, calcium and abscisic acid on anther pretreatment. Plant Sci 126:211–218
36. Cistué L, Vallés MP, Echávarri B et al (2004) Production of barley doubled haploid by anther and microspore culture. In: Mujib A, Myeong-Je C, Predieri S, Banjaree S (eds) *In vitro* applications in crop improvements—recent progress.

Oxford & IBH Publishing Co. Pvt. Ltd, New Delhi, India, pp 1–18

37. Hunter CP (1988) Plant regeneration from microspores of barley, *Hordeum vulgare* L. PhD thesis. Wye College, Univ of London, London
38. Hu TC, Ziauddin A, Simion E et al (1995) Isolated microspore culture of wheat (*Triticum aestivum* L.) in a defined media I. Effects of pretreatment, isolation methods, and hormones. In Vitro Cell Dev Biol 31:79–83
39. Gamborg OL, Miller RA, Ojima K (1968) Nutritional requirements of suspension cultures of soybean root cells. Exp Cell Res 50:151–158
40. Kasha KJ, Simion E, Miner M et al (2003) Haploid wheat isolated microspore culture protocol. In: Malszynnski M, Kasha KJ, Forster BP, Szarejko I (eds) Doubled haploid production in crop plants: a manual. Kluwer Academic Publishers, Dordrecht, pp 77–82
41. Soriano M, Cistué L, Castillo AM (2008) Enhanced induction of microspore embryogenesis after *n*-butanol treatment in wheat (*Triticum aestivum* L.) anther culture. Plant Cell Rep 27:805–811

Chapter 15

Biolistic Transformation of Wheat with Centrophenoxine as a Synthetic Auxin

Ainur Ismagul, Gulnur Iskakova, John C. Harris, and Serik Eliby

Abstract

Cereal crops, including bread wheat (*Triticum aestivum* L.), are an important staple food worldwide. With a growing global population, it is evident that current crop production will not meet the rising demands being placed on modern agriculture. Efforts to improve crop yield and stress-tolerance by traditional breeding are labor intensive, time consuming, and highly dependent upon the ability to capture existing and novel genetic variation from a restricted genetic pool. Genetic engineering of crop species is one of several alternatives to traditional breeding for the introduction of novel genetic variation. This recently established technology has proved useful for the introduction of novel traits like pest resistance and herbicide tolerance. As a universal tool for genetic transformation, the Biolistic Gene Gun allows for the genomic integration of novel gene sequences from various sources into a whole host of living organisms.

In this chapter, we present a novel and detailed protocol for the Biolistic Transformation of bread wheat that uses the pharmaceutical compound, Centrophenoxine (CPX). The application of CPX as the main auxin-like plant growth regulator in cereal genetic transformation replaces the potent but more toxic herbicide 2,4-D.

Key words Biolistic, Cereals, Centrophenoxine, Microprojectile bombardment, Transgenic, Wheat

1 Introduction

The genetic transformation of cereal crops, including wheat, is among the most important and challenging objectives of modern plant biotechnology. Despite the considerable efforts made by the international research community, wheat is the last major crop to be genetically transformed [1]. Biolistic transformation allowed production of the first transgenic wheat, as well as many other transgenic crops including millet, maize, and sorghum, which have been recalcitrant to other methods of transformation [2–4].

Centrophenoxine (CPX, Meclofenoxate) is shown to reduce the development of the ageing associated pigment lipofuscin from neuron cells [5]. CPX is composed of two chemicals: dimethylami noethanol (DMAE), a compound found naturally in food,

Delphine Fleury and Ryan Whitford (eds.), *Crop Breeding: Methods and Protocols*, Methods in Molecular Biology, vol. 1145, DOI 10.1007/978-1-4939-0446-4_15,

especially fish, and *para*-chlorophenoxyacetate (pCPA), a synthetic compound similar to the auxin group of plant hormones [6].

Earlier it was shown that CPX can induce somatic embryogenesis in dicot plant species, including peanuts [7], peppers [8], and peaches [9]. We have found that CPX can be used in cereal tissue culture as a safe alternative to the popular yet more toxic 2,4-D [10]. We have used this protocol, with minor modifications, for the transformation of millet, maize, sorghum, and Eastern Gamagrass (*see* **Note 1**).

2 Materials

2.1 Growth of Wheat Plants

The Australian commercial spring wheat cultivars Gladius, Drysdale, Krichauff, Rees, Westonia, and Yitpi have been genetically transformed using this protocol.

1. Pots: 0.55 L MK12 Masrac Taglok punnet pots (90 × 80 mm), 6 L planting pots (8 in. diameter).
2. Soil Mix: 75 % (v/v) Coco Peat, 25 % (v/v) nursery cutting sand (sharp), 750 mg/L $CaSO_4 \cdot 2H_2O$ (gypsum) 750 mg/L $Ca(H_2PO_4) \cdot 2H_2O$ (superphosphate), 1.9 g/L $FeSO_4$, 125 mg/L FeEDTA, 1.9 g/L $Ca(NO_3)_2$, 750 mg/L Scotts Micromax micronutrients, and 2.5 g/L Osmocote Plus slow release fertilizer (16:3:9) (Scotts Australia Pty. Ltd.). pH is adjusted to between 6.0 and 6.5 using two parts agricultural lime to one part hydrated lime.
3. TPS TOP—clear plastic micropropagating cover (Thermofilm, Springvale, Australia).

2.2 Sterilizing Wheat Caryopses

1. Aqueous ethanol 70 % (v/v), Sodium hypochlorite 4.0 % (m/v), Polyoxyethylenesorbitan monolaurate (Tween 20), Sterile water.
2. Glass beakers, sterile metal sieves.
3. Magnetic stirrer and magnetic spin bar.
4. Austraseal sealing tape (Austratec).

2.3 Culture, Selection, and Regeneration Media

1. Medium for culture of immature embryos (MSC; *see* Table 1 and **Note 2**): Murashige and Skoog (MS; *see* ref. 11) basal medium with vitamins (4.43 g/L), sucrose (30 g/L), casein acid hydrolysate (0.5 g/L), centrophenoxine hydrochloride (CPX, 150 μL/L of stock solution, *see* Table 2). Petri dishes (90 × 25 mm).
2. Medium for osmotic treatment of embryos (MSOS, *see* Table 1 and **Note 3**): MS basal medium with vitamins (4.43 g/L), sucrose (120 g/L), $CuSO_4$ (40 μL/L of stock solution, *see* Table 2), agar (9 g/L). Petri dishes (60 × 15 mm).

Table 1
Composition of culture media (per 1 L)

	Component					
	1	2	3	4	5	6
Type of medium	**MSC**	**MSOS**	**MSC2**	**MSSel**	**MSR**	**½MS**
Murashige–Skoog basal medium with vitamins (1 pk = 4.43 g/L)	1 pk	1 pk	1 pk	1 pk	–	0.5 pk
MS modified basal salt mixture without ammonium nitrate (1 pk = 2.68 g/L)	–	–	–	–	1 pk	–
MS vitamin solution (1,000×)	–	–	–	–	1,000 μL	–
Sucrose	30 g	100 g	60 g	30 g	–	15 g
Casein hydrolysate	0.5 g	0.5 g	1.0 g	0.5 g	1.0 g	0.25 g
$CuSO_4$ $5H_2O$ (25 mM)	40 μL	40 μL	40 μL	40 μL	200 μL	25 μL
Centrophenoxine (200 mg/mL)	150 μL	–	25 μL	25 μL	–	–
Maltose × H_2O	–	–	–	–	20 g	–
Kinetin (1 mg/mL)	–	–	–	–	100 μL	–
Gelrite	5 g	–	5 g	5 g	5 g	4 g
Agar	–	9 g	–	–	–	–
Add after autoclaving:						
Zeatin (5 mg/mL)	–	–	–	–	1,000 μL	–
Hygromycin B (100 mg/mL)	–			1,000 μL	–	500 μL

1. Medium for culture of immature embryos
2. Medium for osmotic treatment of embryos
3. Medium for callus culture
4. Medium for selection
5. Medium for regeneration
6. Medium for root elongation

Table 2
Stock solutions

Hormone/chemical	Concentration	Solvent	Storage
Centrophenoxine hydrochloride (CPX)	200 mg/mL	DMSO	–20 °C, 1 mL aliquots
Kinetin	1 mg/mL	DMSO	–20 °C, 1 mL aliquots
Zeatin	5 mg/mL	DMSO	–20 °C, 1 mL aliquots
MS Vitamin Solution 1,000×	1 mL/L	Water	–20 °C, 1 mL aliquots
Hygromycin B	100 mg/mL	Water	+4 °C
Spermidine	0.1 M	Water	–20 °C, 0.1 mL aliquots
$CaCl_2$	2.5 M	Water	–20 °C, 1 mL aliquots
$CuSO_4 \times 5H_2O$	25 mM	Water	+4 °C

3. Medium for callus culture (MSC2, *see* Table 1): MS basal medium with vitamins (4.43 g/L), sucrose (60 g/L), casein hydrolysate (1 g/L), CPX (25 μL/L), $CuSO_4$ (40 μL/L, *see* Table 2), Gelrite 5 g/L. Petri dishes (90×25 mm).
4. Medium for selection (MSSel, *see* Table 1 and **Note 4**): MS basal medium with vitamins (4.43 g/L), sucrose (30 g/L), casein hydrolysate (0.5 g/L) CPX (25 μL/L), $CuSO_4$ (40 μL/L, *see* Table 2), Gelrite (5 g/L). Petri dishes (90×25 mm).
5. Medium for regeneration (MSR, *see* Table 1 and **Note 5**): MS modified basal salt mixture without ammonium nitrate (2.68 g/L), MS vitamin solution 1,000× (1 ml/L, *see* Table 2), maltose (20 g/L), casein hydrolysate (1 g/L), $CuSO_4$ (200 μL/L, *see* Table 2), zeatin (1 mL/L, *see* Table 2) kinetin (100 μL/L, *see* Table 2), Gelrite (5 g/L). Petri dishes (90×25 mm).
6. Medium for root elongation (½MS, *see* Table 1 and **Note 6**): MS basal medium with vitamins (2.215 g/L), sucrose (15 g/L), casein hydrolysate (0.25 g/L), $CuSO_4$ (25 μL/L, *see* Table 2), Gelrite (2.5 g/L). Petri dishes (90×25 mm). Magenta boxes (75×75×100 mm).

2.4 Particle Bombardment

1. The Biolistic Particle Delivery System PDS-1000/He (Bio-Rad, Hercules, CA, USA).
2. Vacuum pump (Model RV3; Edwards, West Sussex, England).
3. Macrocarriers, stopping screens, 900 or 1,100 psi rupture disks (Bio-Rad).
4. Gold particles: 0.6 μm (Bio-Rad).
5. 2.5 M $CaCl_2$, 0.1 M spermidine (*see* Table 2).
6. Plasmid DNA: prepare using Qiagen Mini-prep kit (Qiagen GmbH, Hilden, Germany).
7. Minimal expression units (MEU): gel purify (*see* Subheading 3.2) using GenElute Extraction Kit (Sigma Chemical Company, St. Louis, MO, USA).
8. TE buffer: 10 mM Tris pH 8.0 with HCl, 1 mM EDTA.
9. TAE buffer: 40 mM Tris pH 8.0 with glacial acetic acid, 1 mM EDTA.
10. NaAc: 3 M sodium acetate pH 4.8.
11. Ethanol solutions at 70 and 100 %.
12. DNA LoBind Eppendorf 1.5 ml tubes (Eppendorf AG, Hamburg, Germany).
13. 10× Loading buffer (Ficoll): 20 % Ficoll, 100 mM EDTA, 1.0 % SDS, 0.25 % bromophenol blue, 0.25 % xylene cyanol.
14. Benchtop microcentrifuge.

2.5 PCR Analyses

1. MJ Research PTC100 Thermocycler (Bio-Rad).
2. Transgene specific oligonucleotide PCR primers and standard PCR reaction reagents.

3 Methods

All procedures are carried out at room temperature unless otherwise specified.

3.1 Growth of Plants for Explant Donor Material

The quality of donor plants is critical to the success of transformation. In order to provide suitable wheat donor material of consistent quality, plants should be grown under environmentally controlled glasshouse conditions (*see* **Note 7**). From March to mid-October (start of Autumn to mid-Spring in Southern Hemisphere), wheat plants are sown 5–6 plants per 6 L (8 in. diameter) pot containing soil mix. For the first 5–6 weeks, grow seedlings under natural lighting conditions as determined by the prevailing season. Plants can then be supplemented with artificial light to extend the day length. Additional light can be provided using 400 W high pressure sodium lamps in combination with metal halide lamps. Additional lighting is predominantly used from mid-April to mid-September and should be adapted to the prevailing seasonal conditions. Generally, supplemented lighting is initiated at 7 a.m. through until 10 a.m. and then again from 4 p.m. to 7 p.m., this extends day length to 12 h.

Ideally, keep growth rooms at a constant 23 °C during the day through the use of air-conditioning. When using evaporative air-conditioning, it can result in small temperature fluctuations (+5 °C) on exceptionally hot and humid days. In order to prevent cooling below 12 °C, use a self-inactivating (at 14 °C) central hot water heating system.

To increase plant tillering during longer summer days (mid-October through mid-February), germinate 5–6 seeds in 6 L (8 in. diameter) pots containing soil mix. Place in growth chambers for 7–8 weeks under a 12 h lighting regime (using metal halide lamps, 400 W) with 20/15 °C day/night temperature. After this period transfer plantlets to environmentally controlled glasshouse conditions.

All donor plant material should be grown in the glasshouse for a period of 3–4 months until immature seed set. Harvest immature seeds suitable for embryo isolation 11–14 days post pollination.

3.2 Immature Wheat Seed Sterilization and Embryo Isolation

1. Place immature wheat seeds in a glass beaker with 300 mL of 70 % ethanol, agitate using a spin bar on a magnetic stirrer for 5 min and remove the ethanol. Follow this by agitation in 300 mL of 1 % sodium hypochlorite solution supplemented with 0.5 μL/mL Tween 20 for 15 min. Remove this solution

and rinse three times with fresh sterile distilled water, 5 min each time (*see* **Note 8**).

2. Isolate immature embryos (*see* **Note 9**) aseptically and place scutellum side up on MSC medium (*see* Table 1).

3.3 Plasmid Preparation and DNA-Gold Coating

All plasmid constructs are purified by alkaline lysis. Minimal expression units (MEU, *see* **Note 10**) are released from the plasmid by digestion with appropriate restriction enzymes:

1. Digest 1–2 μg of plasmid DNA with the appropriate restriction enzymes according to manufacturer's instructions.
2. Prepare a 0.8 % agarose gel in TAE buffer.
3. Load the digested DNA in 15 μL of TE buffer supplemented with 1.5 μL of 10× loading buffer (Ficoll).
4. Electrophorese sample at 25 mA until the dye front has reached the bottom of the gel or the desired DNA fragment containing the MEU is clearly separated from backbone plasmid DNA.
5. After electrophoresis carefully excise a gel slice containing only the desired DNA fragment using a clean and sterile sharp scalpel blade and place in a 2 mL microcentrifuge tube.
6. DNA is purified from the agarose gel slice using the GenElute gel extraction kit (Sigma) and eluted from the column in TE buffer according to the manufacturer's instructions (*see* **Note 11**).
7. Precipitate the eluted DNA by addition of a 1/10 volume of NaAc and 2.5× volume of 100 % ethanol. Centrifuge the precipitated DNA at maximum speed in a benchtop microcentrifuge for 10 min then carefully remove and discard the supernatant. Wash the DNA pellet with 70 % ethanol, centrifuge 5 min at maximum speed, and remove the supernatant. Leave the DNA to air dry so that the white pellet turns colorless, redissolve DNA in 10 μL of sterile Nanopure water.

Prepare DNA-gold coating (*see* **Note 12**) according to Bio-Rad protocol (*see* ref. 12) with the following modifications:

1. Mix 10 μL DNA (0.1–0.2 μg/μL) with 50 μL of gold particles (0.6 μm, 30 mg/mL) in 50 % glycerol.
2. Add 50 μL 2.5 M $CaCl_2$ and 20 μL of 0.1 M spermidine (*see* **Note 13**).
3. Vortex the mixture at 93 × *g* for 2 min, followed by incubation for 20 min at room temperature.
4. Centrifuge at 3,921 × *g* for 1 min, discard supernatant.
5. Wash pellet with 1 mL of 70 % followed by 1 mL of 99.5 % ethanol.
6. Resuspend the pellet in 60 μL of 99.5 % ethanol (*see* **Note 14**).
7. Use 5 μL of DNA–gold suspension per bombardment.

3.4 Bombardment

Microprojectile bombardment is performed according to the standard procedures (*see* refs. 13–16) with the following adaptations:

1. On the day of bombardment the most responsive and precultured embryos are selected under a stereomicroscope. These are identified by cell proliferation on the scutellum or the appearance of embryogenic calli (*see* **Note 15**). Place these tissues on MSOS osmotic medium (*see* Table 1) in the center of a 60 mm petri dish forming a circle of 25–30 mm in diameter. Leave embryonic tissues for a minimum 4 h (*see* **Note 16**).
2. Surface-sterilize the PDS-1000/He device with 70 % (v/v) ethanol.
3. Vortex the DNA coated gold particles to disperse and whilst mixing vigorously with a pipette, remove a 5 μL aliquot and load centrally onto the macrocarrier membrane. Allow all ethanol to evaporate.
4. Conduct biolistic operations according to the Bio-Rad protocol (*see* **Note 17**).
5. Incubate cultures in the dark at room temperature for 16 h, then transfer treated embryos to MSC2 medium (*see* Table 1) for a period of 1 week.

3.5 Selection, Regeneration

1. One week after the bombardment, transfer the treated embryos developing calli to MSSel medium with 100 mg/L Hygromycin B (*see* Table 1) and culture in the dark (*see* **Note 18**).
2. After eight weeks, transfer plates to 16/8 h day/night low light at 24 °C for 3–4 days. Subculture the greening callus tissue to MSR medium (*see* Table 1) without Hygromycin B.
3. Transfer regenerating plantlets to petri plates (90 × 25 mm) containing half-strength hormone-free ½MS medium (*see* Table 1) supplemented with 50 mg/L Hygromycin B.
4. Plantlets exhibiting strong root growth on selection are considered transgenic and subsequently transferred to Magenta boxes containing hormone-free ½MS medium without selection and grown for a further 10 days (*see* **Note 19**).
5. PCR can be used at this point to determine presence of either the MEU or selectable marker gene sequences (e.g., *hpt*) (*see* **Note 20**).
6. Once plantlets have developed sufficient root biomass to support shoot growth they are transplanted to soil in 0.55 L MK12 Masrac Taglok punnet pots (90 × 80 mm) and grown under greenhouse conditions to maturity (*see* **Note 21**).

4 Notes

1. Immature embryos of maize (*Zea mays* L.), millet (*Pennisetum glaucum* L.), and sorghum (*Sorghum bicolor* L.) have been successfully used as transformation explants. For millet and Eastern Gamagrass (*Tripascum dactyloides* L.) shoot apices are used to generate material for transformation. Shoot apices are isolated from 4–5-day-old seedlings, germinated aseptically on hormone-free ½MS medium (*see* Table 1) and cultured on MSC medium (*see* Table 1) for 2–3 weeks. The resulting embryogenic calli serves as explant tissues. Transformation frequencies using these tissues are variable but positively correlate with regeneration capacity of tested genotypes.
2. For media preparation: Adjust pH to 5.8–5.85 with 1 N KOH and bring volume to 1 L with Milli-Q purified water. Pour 0.5 L into two 1 L Schott bottles with the appropriate amount of the gelling agent. Autoclave at 121 °C for 20 min. Cool bottles to approx. 60 °C in a water bath incubator, and in a sterile laminar hood pour 25–30 mL/plate.
3. Pour 6 mL/plate after autoclaving.
4. Add 1,000 μL/L Hygromycin B (*see* Table 2) once media has cooled to 60 °C in water bath after autoclaving. In a sterile laminar hood pour 25–30 mL/plate.
5. After autoclaving, cool media to 60 °C in water bath and add 1,000 μL/L Zeatin (*see* Table 2) and pour 25–30 mL/plate.
6. After autoclaving, cool media to 60 °C in water bath and add 500 μL/L Hygromycin B (*see* Table 2) and pour 25–30 mL per plate or Magenta Box.
7. Care should be taken to maintain healthy, unstressed plants in a pest free environment. Additional lighting should be used during the winter season to extend day length to 12 h.
8. Careful treatment and peeling of the immature seeds is important for the isolation of undamaged embryos. Size of beaker, speed of rotation on the stirrer, size of magnetic bar and volumes of solutions can be adjusted according to the number of immature seeds from which embryos are to be isolated. Important—all seeds must be in contact with the solution and easily suspended.
9. MEU contains promoter-coding sequence-terminator.
10. It is very important to isolate intact unbroken embryos. Use sharp blades and place the isolated embryos onto the surface of the cultivation medium at once. After having isolated 50–70 per plate, arrange embryos scutellum side up. Seal the plates with Austraseal tape.

11. To improve yield: place the excised agarose piece with the DNA band in a 2 mL Eppendorf tube and freeze at −80 °C for 5–10 min. Preheat the elution buffer to 65 °C before use.
12. Prior to use, centrifuge the DNA solution at maximum speed for 5–10 min to sediment particulate matter. Prepare 50 μL aliquots of gold in DNA LoBind Eppendorf tubes to prevent "sticking."
13. Spermidine solutions are unstable—prepare fresh 0.1 mL aliquots, discard excess.
14. Gold particle agglomeration is detrimental to the efficiency of transformation. The DNA–gold mixture should readily resuspend after the final centrifugation step. If agglomerates form, repeat DNA-gold coating using less DNA. Using high-quality DNA dissolved in ultrapure sterile water aids the DNA-gold coating procedure and increases transformation efficiency. DO NOT overload gold particles with DNA (<5 μg/tube). Use aliquots of absolute ethanol.
15. Selection of embryos is crucial and active callus production must be identified. Embryos as young as 4–5 days may be selected and approximately 100 will be required to fill a circle of 25–30 mm diameter. At 2 weeks of incubation on MSC media, active embryos will appear larger and may require only 50–60 embryos to fill the 25–30 mm diameter. If it is difficult to distinguish at this stage, embryos can be cultured for up to three weeks but transformation efficiency will decline.
16. Osmotic pretreatment can be done overnight if needed.
17. The PDS-1000/He device is used in conjunction with a cylinder of compressed helium. Refer to Chemical Management and Safety Policies and appropriate Material Safety Data Sheets (MSDS).

    Bombardment Procedure (for more details refer to the Bio-Rad manual):

    (a) Turn the PDS-1000/He device power ON.
    (b) Slowly open the main valve of the helium cylinder until gas pressure registers on the first gauge of the regulator (gauge closest to the gas cylinder). Turn the regulator adjusting-handle clockwise to ensure the helium tank pressure regulator is set 200 psi over the selected rupture disk pressure.
    (c) Turn ON vacuum pump.
    (d) Load rupture disk holder, microcarrier launch assembly and the petri dish containing the target tissue.
    (e) Set the VACUUM switch to the VAC position to evacuate the sample chamber. Use a vacuum of 26–28 in. Hg.

(f) Once vacuum is reached, press the VACUUM switch to the HOLD position.

(g) Press and hold the FIRE switch to allow helium pressure to increase in the accelerator tube. The gauge at the top of the accelerator tube should be observed. The rupture disk should burst at the selected rupture pressure.

(h) Release the FIRE switch immediately after the rupture disk bursts.

(i) Release the vacuum in the sample chamber by setting the VACUUM switch to the VENT position.

(j) After releasing the vacuum, open the chamber door and remove the petri dish.

(k) Remove the macrocarrier launch assembly and discard the macrocarrier and the stopping screen.

(l) Unscrew the rupture disk holder and discard the remains of the rupture disk.

(m) If the PDS-1000/He device is not used again, close the main valve on the helium cylinder.

(n) Without installing a rupture disk, draw a vacuum until the FIRE switch is illuminated.

(o) Press and hold the FIRE switch until both pressure gauges on the regulator read "0."

(p) Release the tension on the pressure-adjusting screw by turning the handle anticlockwise until it turns freely.

(q) Once the tension is released, set the VACUUM switch to the VENT position, and the instrument can be shut off.

(r) Turn OFF vacuum pump.

The distance from the macrocarrier launch point to the stopping screen should be set at 15 mm, while a distance of 60 mm should be between the stopping screen and target tissue. The distance between the rupture disk and launch point of the macrocarrier is 30 mm.

18. Hygromycin selection (*see* ref. 17): When strong promoters like maize *Ubiquitin-1* are used, improved selection is possible on higher concentrations of 50–100 mg/L of the antibiotic. In wheat the *CaMV35S* promoter is not strong enough to provide efficient selection. Many selectable markers can be adapted for use in cereal genetic transformation (*see* Table 3).

19. Vigorously growing roots are indicative of positive selection, however for some transgenic plantlets this may not be evident. Therefore a second round of selection for a further 10 days on hormone-free ½MS medium with 50 mg/L Hygromycin B is suggested to increase selection efficiency.

Table 3
Selectable markers in wheat genetic transformation

Gene	Enzyme	Origin	Selection	Mechanism of action
hpt, *hph*, or *aph IV*	Hygromycin phosphotransferase	*E. coli*	Hygromycin B (50–150 mg/L)	Phosphorylation
bar, *pat*	Phosphinothricin-N-acetyltransferase	*Streptomyces hygroscopicus* and *S. viridochromogenes*	Bialaphos, phosphinothricin (5–10 mg/L)	Acetylation
aroA:CP4, *epsps*	5-Enol-pyruvylshikimate-3-phosphate synthase	*Agrobacterium* sp. CP4 or *Salmonella typhimurium*	Glyphosate (0.1–2.0 mM)	Glyphosate-insensitive enzyme (in chloroplasts)
cah	Cyanamide hydratase	*Myrothecium verrucaria*	Cyanamide (37.5 mg/L)	Converts cyanamide to urea
pmi, *manA*	Phosphomannose isomerase	*E. coli*	Mannose (5–10 g/L)	Converts mannose-6-phosphate into fructose-6-phosphate
als, *csr1*, *ahas*	Acetolactate synthase (acetohydroxy acid synthase)	*Oryza sativa* L., *Triticum aestivum* L.	Herbicide (0.1–2.0 mM)	Sulfonylurea, imidazolinone-resistant enzymes

20. We use the REDextract-N-Amp Plant PCR Kit (Sigma), which has been adapted for smaller volumes as follows: 4–5 mm of leaf tip is cut from the putative transgenic and placed in a 200 μL PCR tube with 40 μL of extraction buffer and left on ice until all samples are collected. Samples are then heated to 95 °C for 10 min and briefly cooled to 15 °C. 40 μL of dilution buffer is then added and samples briefly vortexed. This solution is used as a template for transgene detection PCR. Depending upon the DNA sequence(s) used for transformation, oligonucleotide primers specific to the transgene should be designed and used for PCR. For wheat transformation, the *hpt* selectable marker sequence is used routinely for transgene detection. Primers specific for the *hpt* coding sequence used for amplification are as follows:

    hptF: 5′-agacctgcctgaaaccgaactgc-3′; hptR: 5′-cacggcctccagaagaagatgttg-3′; PCR fragment size: 411 bps.

21. Carefully remove the plantlets from the jars. Using sharp scissors cut off any necrotic leaves or calli. Wash out the roots in tap water to remove any remaining Gelrite. Holding the plantlets flat on your hand, gently transplant them into very moist soil in 0.55 L MK12 Masrac Taglok punnet pots (90 × 80 mm) and cover roots completely. Do not push roots

into the soil as this risks damage. Use a plastic cover for 3–5 days to create a humid environment and a tray with wetted paper towels on the bottom large enough for 12× 0.55 L MK12 pots. We use a TPS TOP with a 3-way vent system to allow a controlled decrease in the humidity of the chamber and a gradual acclimation to greenhouse conditions.

Acknowledgments

We thank Prof. T. Shimada (The Ishikawa Agricultural College, Japan), Drs. R. Kandzia and M. Rubtsova (Icon Genetics, Germany) for their contribution to the development of the present protocol, and Dr. U. Langridge for her contribution to plant growth methodologies.

References

1. Vasil V, Castillo AM, Fromm ME et al (1992) Herbicide resistant fertile transgenic wheat plants obtained by microprojectile bombardment of regenerable embryogenic callus. Bio/Technol 10:667–674
2. Lambe P, Dinant M, Deltour R (2000) Transgenic pearl millet (*Pennisetum glaucum*). In: Bajaj YPS (ed) Transgenic crops, vol 46, Biotechnology in agriculture and forestry. Springer, Berlin, pp 84–108
3. Klein TM, Fromm ME, Weissinger A et al (1988) Transfer of foreign gene into intact maize cells with high-velocity microprojectiles. Proc Natl Acad Sci U S A 85:4305–4309
4. Casas AM, Kononowicz AK, Zehr UB et al (1993) Transgenic plants via microprojectile bombardment. Proc Natl Acad Sci U S A 90:11212–11216
5. Zhang L, Hui YN, Wang YS et al (2011) Calcium overload is associated with lipofuscin formation in human retinal pigment epithelial cells fed with photoreceptor outer segments. Eye (Lond) 25:519–527
6. Nehra B, Verma R, Khanna P et al (2008) Behavioural alterations in rotenone model of Parkinson's disease: attenuation by co-treatment of Centrophenoxine. Brain Res 1201:122–127
7. Little EL, Magbanua ZV, Parrott WA (2000) A protocol for repetitive somatic embryogenesis from mature peanut epicotyls. Plant Cell Rep 19:351–357
8. Steintiz B, Küsek M, Tabib Y et al (2003) Pepper (*Capsicum annuum* L.) regenerants obtained by direct somatic embryogenesis fail to develop a shoot. In Vitro Cell Dev Biol-Plant 39:296–303
9. Srinivasan C, Scorza R (2006) The influence of genotype on the induction of somatic embryos from *in vitro* cultured zygotic embryos of peach. Acta Horticult 738:691–696
10. Bus JS, Hammond LE (2007) Regulatory progress, toxicology, and public concerns with 2,4-D: where do we stand after two decades? Crop Prot 26:266–269
11. Murashige T, Skoog F (1962) A revised medium for rapid growth and bioassay with tobacco tissue culture. Physiol Plant 15:473–497
12. Sanford JC, Smith FD, Russel JA (1993) Optimizing the biolistic process for different biological applications. In: Wu R (ed) Methods in enzymology. Elsevier Inc., New York, pp 483–509
13. Shimada T, Seki M, Morikawa H (1991) Transient expression of *beta-glucuronidase* (GUS) gene in wheat pollen embryos via microprojectile bombardment. Wheat Inf Serv 72:106–108
14. Vasil V, Srivastava V, Castillo AM et al (1993) Rapid production of transgenic wheat plants by direct bombardment of cultured immature embryos. Bio-Technol 11:1553–1558
15. Rubtsova M, Kempe A, Ismagul A et al (2008) Expression of active *Streptomyces* phage *phiC31* integrase in transgenic wheat plants. Plant Cell Rep 27:1821–1831
16. Morran S, Eini O, Pyvovarenko T et al (2011) Improvement of stress tolerance of wheat and barley by modulation of expression of DREB/CBF factors. Plant Biotechnol J 9:230–249
17. Ortiz JPA, Reggiardo MI, Ravizzini RA et al (1996) Hygromycin resistance as an efficient selectable marker for wheat stable transformation. Plant Cell Rep 15:877–881

Chapter 16

Agrobacterium-Mediated Transformation of Barley (*Hordeum vulgare* L.)

Ainur Ismagul, Iryna Mazonka, Corinne Callegari, and Serik Eliby

Abstract

Barley biotechnology requires efficient genetic engineering tools for producing transgenic plants necessary for conducting reverse genetics analyses in breeding and functional genomics research. *Agrobacterium-mediated* genetic transformation is an important technique for producing barley transgenics with simple low-copy number transgenes. This chapter reports a refined protocol for the systematic high-throughput transformation of the advanced Australian spring barley breeding line WI4330.

Key words *Agrobacterium*, Barley, Biotechnology, Cereals, Transgenic

1 Introduction

Barley is one of the most ancient and environmentally adaptive domesticated cereal crops. It is the fourth most important cereal according to production levels worldwide and is the second most important crop following wheat in Australia [1, 2]. The first fertile transgenic barley plants were produced using the biolistic transformation method in 1994 [3]. Tingey and coworkers, shortly thereafter transformed barley using *Agrobacterium*-mediated infection of immature embryos [4]. Over the last decade there have been a number of studies reporting on the generation of barley transgenics. Most of these studies rely either on the Scottish spring-type cultivar Golden Promise or the German winter-type cultivar Igri, as they readily form callus and regenerate in vitro [5–7]. Neither of these genotypes has agronomic properties suitable for a winter crop in typical Mediterranean type climatic environments (cool wet winters, hot dry summers) which is experienced throughout most of Australia's barley growing regions. Therefore, this chapter presents a modified *Agrobacterium*-mediated transformation protocol that has routinely been used and resulted in the production of over one thousand independent transgenic events of the Australian-adapted barley breeding line WI4330 [8].

Delphine Fleury and Ryan Whitford (eds.), *Crop Breeding: Methods and Protocols*, Methods in Molecular Biology, vol. 1145, DOI 10.1007/978-1-4939-0446-4_16, © Springer Science+Business Media New York 2014

2 Materials

2.1 Growth of Barley Plants

Australian advanced barley breeding line WI4330 is a doubled haploid breeding line from the cross Maritime/Commander// Flagship and has been genetically transformed using this protocol.

1. Pots: 0.55 L MK12 Masrac Taglok punnet pots (90 × 80 mm), 6 L planting pots (8 in. diameter).
2. Soil Mix: 75 % (v/v) Coco Peat, 25 % (v/v) nursery cutting sand (sharp), 750 mg/L $CaSO_4{\cdot}2H_2O$ (gypsum) 750 mg/L $Ca(H_2PO_4){\cdot}2H_2O$ (superphosphate), 1.9 g/L $FeSO_4$, 125 mg/L FeEDTA, 1.9 g/L $Ca(NO_3)_2$, 750 mg/L Scotts Micromax micronutrients, and 2.5 g/L Osmocote Plus slow release fertilizer (16:3:9) (Scotts Australia Pty. Ltd.). pH is adjusted to between 6.0 and 6.5 using two parts agricultural lime to one part hydrated lime.
3. TPS TOP—clear plastic micropropagating cover (Thermofilm, Springvale, Australia).

2.2 Sterilization of Barley Seeds and Isolation of Immature Embryos

1. Aqueous ethanol 70 % (v/v), Sodium hypochlorite solution 30 % (v/v), sterile deionized water.
2. Glass beakers.
3. Automated pipette and sterile pipettes.
4. Microforceps and scalpels.
5. Dissecting stereomicroscope.

2.3 Culture Media

1. Co-cultivation medium (MSC; *see* Table 1 and **Note 1**): Murashige and Skoog (MS; *see* ref. 9) basal medium with vitamins (4.43 g/L), maltose $\times H_2O$ (30 g/L), casein hydrolysate (1.0 g/L), proline (0.69 g/L), inositol (0.25 g/L), dicamba (50 μL/L of stock solution, *see* Table 2), Gelrite (5 g/L). Use petri dishes (90 × 25 mm).
2. Recovering medium (MSRc, *see* Table 1): MS basal medium with vitamins (4.43 g/L), sucrose (30 g/L), casein hydrolysate (1.0 g/L), $CuSO_4 \times 5H_2O$ (40 μL/L of stock solution, *see* Table 2), proline (0.69 g/L), inositol (0.25 g/L), dicamba (50 μL/L, *see* Table 2), Gelrite (5 g/L), timentin (1 mL/L of stock solution, *see* Table 2). Petri dishes (90 × 25 mm).
3. Selection medium (MSSel, *see* Table 1): MS basal medium with vitamins (4.43 g/L), sucrose (30 g/L), casein hydrolysate (1.0 g/L), $CuSO_4 \times 5H_2O$ (40 μL/L, *see* Table 2), proline (0.69 g/L), inositol (0.25 g/L), dicamba (50 μL/L, *see* Table 2), Gelrite (5 g/L), timentin (1 mL/L, *see* Table 2), hygromycin B (500 μL/L of stock solution, *see* Table 2). Petri dishes (90 × 25 mm).

Table 1
Composition of culture media (per 1 L)

	Type of media				
	1	2	3	4	5
Component	**MSC**	**MSRc**	**MSSel**	**MSRg**	**½ MS**
Murashige–Skoog basal medium with vitamins (1 pk = 4.43 g/L)	1 pk	1 pk	1 pk	–	0.5 pk
MS modified basal salt mixture without ammonium nitrate (1 pk = 2.68 g/L)	–	–	–	1 pk	–
MS vitamin solution (1,000×)	–	–	–	1,000 μL	–
Sucrose	–	30 g	30 g	–	15 g
Dicamba (25 mg/mL)	50 μL	50 μL	50 μL	–	–
Casein hydrolysate	1,000 mg	1,000 mg	1,000 mg	1,000 mg	250 mg
$CuSO_4 \times 5H_2O$ (25 mM)	–	40 μL	40 μL	200 μL	25 μL
Maltose × H_2O	30 g	–	–	20 g	–
Kinetin (1 mg/mL)	–	–	–	100 μL	–
Proline	690 mg	690 mg	690 mg	–	–
Inositol	250 mg	250 mg	250 mg		–
Gelrite	5 g	5 g	5 g	5 g	4 g
Add after autoclaving:					
Timentin (150 mg/mL)	–	1,000 μL	1,000 μL	1,000 μL	–
Zeatin (5 mg/mL)	–	–	–	1,000 μL	–
Hygromycin B (100 mg/mL)	–	–	500 μL	–	500 μL

1. Co-cultivation medium
2. Recovering medium
3. Selection medium
4. Regeneration medium
5. Root elongation medium

4. Regeneration medium (MSRg, *see* Table 1): MS modified basal salt mixture without Ammonium Nitrate (2.68 g/L), MS Vitamin Solution (1 mL/L of stock solution, *see* Table 2), maltose × H_2O (20 g/L), casein hydrolysate (1 g/L); $CuSO_4 \times 5H_2O$ (200 μL/L, *see* Table 2), kinetin (0.1 mL/L of stock solution, *see* Table 2), Gelrite (5 g/L), zeatin (1 mL/L of stock solution, *see* Table 2), timentin (1 mL/L, *see* Table 2). Petri dishes (90 × 25 mm).
5. Root elongation medium (½MS, *see* Table 1) MS basal medium with vitamins (2.215 g/L), sucrose (15 g/L), casein hydrolysate

Table 2
Stock solutions

Hormone/chemical	Concentration	Solvent	Storage
Dicamba	25 mg/mL	DMSO	–20 °C, 1 mL aliquots
Rifampicin	50 mg/mL	DMSO	–20 °C, 1 mL aliquots
Kinetin	1 mg/mL	DMSO	–20 °C, 1 mL aliquots
Zeatin	5 mg/mL	DMSO	–20 °C, 1 mL aliquots
Biotin	0.1 mg/mL	water	–20 °C, 10 mL aliquots
Kanamycin (FS)	50 mg/mL	water	–20 °C, 1 mL aliquots
Hygromycin B (FS)	100 mg/mL	water	+4 °C
MS vitamin solution 1,000×	1 mL/L	water	–20 °C, 1 mL aliquots
Timentin (FS)	150 mg/mL	water	Freshly made
$CaCl_2$	20 mM	water	+4 °C, 10 mL aliquots
$CuSO_4 \times 5H_2O$	25 mM	water	+4 °C

FS filter sterilized

(0.25 g/L), $CuSO_4 \times 5H_2O$ (25 μL/L, *see* Table 2), Gelrite (4 g/L), hygromycin B (500 μL/L, *see* Table 2). Petri dishes (90 × 25 mm). Magenta boxes (75 × 75 × 100 mm).

2.4 Agrobacterium Cultivation Media

1. LB: Bacto Tryptone (10 g/L), yeast extract (5 g/L), NaCl (5 g/L), agar (15 g/L). Adjust pH to 7.5 with 0.1 N NaOH before autoclaving.
2. MG\L: Mannitol (5 g/L), L-glutamic acid (1 g/L), KH_2PO_4 (0.25 g/L), NaCl (0.1 g/L), $MgSO_4 \times 7H_2O$ (0.1 g/L), biotin (10 mL/L of stock solution, *see* Table 2), Bacto Tryptone (5 g/L), yeast extract (2.5 g/L). Adjust pH to 7.0 with 0.1 N NaOH before autoclaving.
3. TYNG: Bacto tryptone (10 g/L), yeast extract (5 g/L), NaCl (5 g/L), $MgSO_4 \times 7H_2O$ (0. 2 g/L). Adjust pH to 7.5 with 0.1 N NaOH before autoclaving.

   Add antibiotics to a concentration of 50 mg/L after autoclaving (Rifampicin, Kanamycin, *see* Table 2, **Note 2**).
4. *Agrobacterium tumefaciens* strains AGL0 or AGL1.

2.5 PCR Analyses

1. MJ Research PTC100 Thermocycler (Bio-Rad).
2. Transgene specific oligonucleotide PCR primers and standard PCR reaction reagents.

2.6 Equipment

1. Spectrophotometer.
2. Benchtop microcentrifuge.

3 Methods

3.1 Growth of Donor Plants

In order to ensure consistent high-quality transformation donor material, grow barley plants of the cultivar WI4330 under environmentally controlled glasshouse conditions. From the middle of February through to the end of October (end of summer till mid-Spring in Southern Hemisphere), barley plants can be sown 5–6 plants per 6 L planting pot (8 in. diameter) containing soil mix.

For the first 5–6 weeks, grow seedlings under natural lighting conditions as determined by the prevailing season. Plants should then be supplemented with artificial light to extend the day length. Additional light can be provided using 400 W high pressure sodium lamps in combination with metal halide lamps. Additional lighting is predominantly used from mid-April to mid-September and should be adapted to the prevailing seasonal conditions. Generally, supplemented lighting is initiated at 7 a.m. through until 10 a.m. and then again from 4 p.m. to 7 p.m., this extends day length to 12 h.

Ideally, keep growth rooms at a constant 23 °C during the day through the use of air-conditioning. When using evaporative air-conditioning, it can result in small temperature fluctuations (+5 °C) on exceptionally hot and humid days. In order to prevent cooling below 12 °C, use a self-inactivating (at 14 °C) central hot water heating system.

All donor plant material is grown in the glasshouse for a period of 3–4 months until immature seed set. Immature seeds suitable for embryo isolation are harvested 11–14 days post pollination.

3.2 Immature Barley Seed Sterilization and Embryo Isolation

1. Harvest immature barley seeds approximately 14 days after flowering. Embryos 1.5–2 mm in diameter are ideal (*see* **Note 3**).
2. Place immature barley seeds in any sealable sterile container and add enough 70 % (v/v) ethanol so that all seeds are completely submersed. Seal container and mix on a shaker for 5 min, remove ethanol, and add 100 mL of 3 % (v/v) sodium hypochlorite solution. Mix for 20 min before removing the solution and rinsing four times with sterile deionized water (*see* **Note 4**). Remove all water but allow seeds to remain wet.
3. Under a dissecting microscope and using a pair of fine micro-forceps and scalpel, isolate the embryo from the seed and remove the embryonic axis. Place the isolated embryo without the embryogenic axis, scutellum side down on the co-cultivation media (*see* **Note 5**). Place 30 embryos per plate, keeping them

side-by-side in a 5×6 orientation as close as possible to one another. Embryos can be stored at room temperature in the dark until inoculation with *Agrobacterium* (*see* **Note 6**).

3.3 Preparation of Competent Agrobacterium Cells

1. Start an overnight culture of *Agrobacterium* in 5 mL TYNG medium + Rifampicin (1 mL/L of stock solution, *see* Table 2, Rif50).
2. On the following day, add 1 mL of overnight culture to 30 mL TYNG + Rif50 in a 50 mL centrifuge tube.
3. Shake cultures at 28 °C for 4–6 h until slightly cloudy (OD600 at 0.2–0.4).
4. Chill cultures on ice for 10 min.
5. Spin cultures at 4,528 ×*g* for 10 min at 4 °C.
6. Discard supernatant and resuspend each pellet in 500 μL ice-cold 20 mM $CaCl_2$.
7. Combine all the cells into one 50 mL centrifuge tube; dispense 100 μL aliquots into prechilled 1.5 mL Eppendorf tubes. Freeze cells in liquid nitrogen and store at −80 °C for future transformation.

3.4 Freeze–Thaw Transformation of Agrobacterium Culture

1. Thaw cells at room temperature.
2. Add 1 μg of plasmid DNA to cells (*see* **Note** 7), mix gently by swirling pipette tip.
3. Freeze cells in liquid nitrogen for 1 min.
4. Thaw cells for 5 min in a 37 °C water bath.
5. Add 500 μL of TYNG medium and incubate shaking at 28 °C for 0.5–2 h.
6. Plate 150 μL of culture onto plates of LB + Rif50 + selective antibiotic.
7. Incubate at 28 °C—colonies will gradually appear over 2 days.
8. After PCR confirmation of positive transformation, take the selected colony and grow overnight in 1 mL LB media + Rif50 + selective antibiotic.
9. When culture is no longer transparent make aliquots of 30 μL, adding 30 μL of 30 % glycerol.
10. Vortex and place in liquid nitrogen and store at −80 °C.

3.5 Co-cultivation Step

1. Add 20 μL of culture to 5 mL of MG\L nutrient medium and incubate at 28 °C while gently agitating for approximately 16 h.
2. Check culture density using a spectrophotometer at 600 nm absorbance. Adjust culture density to 0.3–0.5 OD600.
3. Make 100 μL aliquots of culture. Use 80 μL for infection of embryos.

3.6 Co-cultivation of Isolated Immature Embryos

1. Taking 80 µL of *Agrobacterium* culture, pipette it onto the surface of the embryos so that each is covered by the culture. Tip the plate so that excess culture drains off the embryos.
2. To further reduce the amount of culture on its surface, gently drag the embryos across the media. Make sure to "wipe" both sides before transferring the embryo to a new plate of co-cultivation media, keeping the embryos scutellum side down (*see* **Note 8**).
3. Store infected embryos in the dark, at room temperature for 3 days.

3.7 Recovery, Selection, and Regeneration

1. Following co-cultivation, transfer embryos to recovery media (*see* Table 1) for 2 weeks and succeeding this, transfer calli to selection media for 4 weeks, changing to fresh media after 2 weeks. During selection keep calli in the dark at room temperature.
2. Transfer calli to regeneration media to stimulate growth of shoots. Maintain calli on regeneration media for 6–9 weeks, transferring to fresh media after every 3 weeks, and grow in growth chambers set at low light, 24 °C, with 8 h light and 16 h dark (*see* **Note 9**).
3. Transfer regenerating plantlets to petri dishes containing half strength hormone-free ½MS media without hygromycin. Use Magenta boxes for those that are too large to be contained in petri dishes (*see* **Note 10**).
4. Transfer plantlets with substantial root development to Magenta boxes with hormone-free ½MS medium containing hygromycin. Allow these plantlets to grow until they have developed sufficient root and shoot biomass to support growth in soil under greenhouse conditions. At this stage transplant into 0.55 L MK12 Masrac Taglok punnet pots containing soil mix and grow to maturity (*see* **Note 11**).
5. Before transfer to glasshouse conditions, extract DNA from leaf samples and use PCR to confirm presence of the selectable marker gene. Only transfer plants positive for this gene.

4 Notes

1. For media preparation: Adjust pH to 5.8–5.85 with 1 N KOH and bring volume to 1 L with Milli-Q water. Pour 0.5 L into two 1 L bottles with the appropriate amount of the gelling agent. Autoclave at 121 °C for 20 min. Cool bottles to approx. 60 °C in a water bath incubator, and in a sterile laminar hood pour 25–30 mL/plate.

2. Types of the selective antibiotics can vary depending on the selective marker genes used in different binary vectors (Kanamycin at 50 mg/L, Carbenicillin at 100 mg/L, Spectinomycin at 50 mg/L, etc.).
3. Immature seed may be harvested prior to the day of embryo isolation. Embryo size can be checked prior to harvest by gently peeling back the top layer and squeezing out the embryo by pressing down on the middle of the seed. When harvesting prior to isolation, cut well below the head and place directly into water. These heads can then be stored at 4 °C for a short time until needed. For optimum results, it is best to use healthy plant material free of disease and heads from primary tillers.
4. The awns must be carefully trimmed from the barley heads before removal and sterilization of seeds using sharp scissors. Sterilization and isolation process must be undertaken aseptically. Volumes of ethanol and sodium hypochlorite solution, along with the size of the beaker and speed of agitation are dependent on the total quantity of seeds to be sterilized. All seeds need to be completely submersed in the solutions and easily suspended.
5. To extract the embryo from the seed, hold the seed firm with forceps and cut the tip off. This will create an opening to the embryo cavity and, by applying gentle pressure to the middle of the seed, allow the embryo to slide out. Once the embryo has been extracted, use the forceps to stabilize it by holding the tip against one side. Insert the scalpel blade into the groove between the scutellum and embryo axis and remove the axis. It is important not to damage the fragile embryos during this process.
6. Embryos may be isolated 1 day prior to inoculation with *Agrobacterium* and stored at room temperature in the dark or can be isolated on the same day as inoculation.
7. The amount of DNA added can be adjusted to be between 0.2 and 2 μg depending on the size of the plasmid and the quality of the competent cells used.
8. It is important to limit the excess *Agrobacterium* in order to prevent overgrowth. This is especially important with smaller embryos. Co-cultivation media is free of antibiotics such as timentin or hygromycin. This allows the *Agrobacterium* to grow and infect the embryos.
9. During transfer of callus, remove dead/brown callus from cultures.
10. Regeneration media without hygromycin is used when the plantlets are still small to encourage growth by eliminating selection pressure. However, it is important to maintain timentin

selection to prevent *Agrobacterium* outgrowth. When transferring plantlets remove callus from base of the plant.

11. When transferring from ½MS medium to soil mix, carefully remove plant from the media and remove as much gel from the roots as possible (this can be done in water). Place into a small 0.55 L MK12 Masrac Taglok punnet pot (90 × 80 mm) containing soil mix. Water well and cover with a TPS TOP—clear plastic micropropagating cover.

Acknowledgment

We thank Dr. Rohan Singh (ARC Centre of Excellence in Plant Cell Walls, Australia) who established the original barley transformation protocol while working at the ACPFG.

References

1. Goedeke S, Hensel G, Kapusi E et al (2007) Transgenic barley in fundamental research and biotechnology. Trans Plant J 1:104–117
2. Fell J, Miller M, Brown A et al (2011) Australian crop report—Australian Natural Resources Data Library. No. 160, ABARES, Canberra. http://adl.brs.gov.au/data/warehouse/aucrpd9abcc003/aucrpd9abcc003201112/ACR11_4_Dec_rev1.0.0.pdf. Accessed 16 Oct 2012
3. Wan Y, Lemaux PG (1994) Generation of large numbers of independently transformed fertile barley plants. Plant Physiol 104:37–48
4. Tingay S, McElroy D, Kalla R et al (1997) *Agrobacterium tumefaciens*-mediated barley transformation. Plant J 11:1369–1376
5. Bartlett JG, Alves SC, Smedley M et al (2008) High-throughput *Agrobacterium*-mediated barley transformation. Plant Methods 4:1–22
6. Hensel G, Kastner C, Oleszczuk S et al (2009) Agrobacterium-mediated gene transfer to cereal crop plants: current protocols for barley, wheat, triticale, and maize. Int J Plant Genomics. ID 835608:9
7. Harwood WA (2012) Advances and remaining challenges in the transformation of barley and wheat. J Exp Bot 63:1791–1798
8. Li M, Singh R, Bazanova N et al (2008) Spatial and temporal expression of endosperm transfer cell-specific promoters in transgenic rice and barley. Plant Biotechnol J 6:465–476
9. Murashige T, Skoog F (1962) A revised medium for rapid growth and bioassay with tobacco tissue culture. Physiol Plant 15:473–497

Chapter 17

qPCR for Quantification of Transgene Expression and Determination of Transgene Copy Number

Stephen J. Fletcher

Abstract

Quantitative real-time PCR (qPCR) is a mature technology that can be used to accurately quantify the number of copies of a target nucleic acid in a sample. Here, we describe a method for using this technology to determine the copy number of a transgene stably integrated into a plant's genome and to ascertain the level of transgene expression.

Key words Quantitative Real-Time PCR, qPCR, Transgene, Copy number, Gene expression, Normalization

1 Introduction

The advent of the *q*uantitative Real-Time *P*olymerase *C*hain *R*eaction (qPCR) allowed for the rapid, specific, and sensitive quantification of target nucleic acid sequences. qPCR has a variety of applications and is particularly useful for quantifying the number of copies of a sequence in an organism's genome and quantifying the expression of a gene of interest. Its large dynamic detection range permits the simultaneous quantification of target sequences of vastly different relative abundance. Here, we describe simple and rapid methods for the quantification of the number of transgene copies integrated into a plant's genome and quantification of mRNA copies transcribed from a transgene of interest. The template upon which qPCRs are performed is DNA; for transgene copy-number determination this template is genomic DNA, and for determination of transgene expression levels, cDNA reverse-transcribed from mRNA is used. Rapid methods for extraction of genomic DNA and total RNA from plant tissue are set out, along with reverse-transcription of mRNA to cDNA for transcription analysis, quantification of transgene expression using SYBR-Green chemistry and determination of transgene copy number

Delphine Fleury and Ryan Whitford (eds.), *Crop Breeding: Methods and Protocols*, Methods in Molecular Biology, vol. 1145, DOI 10.1007/978-1-4939-0446-4_17, © Springer Science+Business Media New York 2014

using TaqMan chemistry. Oligonucleotide primer and TaqMan probe sequences are described for a number of important crop and model species.

2 Materials

Prepare all solutions using ultrapure water and analytical grade reagents. Prepare and store all reagents at room temperature unless indicated otherwise.

Contamination at any stage of the process can lead to inaccurate results. Particular care must be taken to ensure stocks of oligonucleotide primers, TaqMan probes, qPCR Master Mixes, and DNA, RNA, and cDNA samples remain free of contamination from foreign nucleic acids. Care must be taken to exclude environmental RNases when handling RNA samples.

As with all protocols carried out in a laboratory setting, ensure that risk assessments are carried out for the process itself and each individual reagent and Safe Operating Procedures (SOPs) are devised and followed to ensure adequate safety.

2.1 Quantification of Transgene Expression

2.1.1 Total RNA Extraction

1. Accurately calibrated pipettes with appropriated sterile pipette tips.
2. Plastic pestles for crushing leaf tissue in 2 mL microfuge tubes.
3. A fume cupboard.
4. Refrigerated microfuge capable of maintaining 12,000 × *g* and a temperature of 4 °C.
5. Spectrophotometer for RNA quantification (NanoDrop products, Wilmington, DE, USA).
6. Liquid nitrogen.
7. TRIzol® reagent (Life Technologies Corporation, Grand Island, NY, USA). This reagent is highly toxic. Read the MSDS and prepare a Safe Operating Procedure before use, use the correct safety equipment and dispose of properly.
8. Chloroform. This reagent is highly toxic. Read the MSDS and prepare a Safe Operating Procedure before use, use the correct safety equipment and dispose of properly.
9. Isopropanol.
10. 75 % ethanol solution (75 mL ethanol made up to 100 mL with dH_2O).
11. Sterile RNase-free dH_2O for RNA suspension.
12. 2 mL microfuge tubes.
13. 1.5 mL microfuge tubes.

2.1.2 cDNA Synthesis

1. Thermal cycler (PCR machine) for incubating reactions at set temperatures.
2. Ice.
3. High-quality total RNA with a concentration above 150 ng/μL.
4. DNase I, Amplification Grade, with 10× reaction buffer (Ambion®, Life Technologies).
5. 50 μM oligo $(dT)_{18}$ primer or random hexamers.
6. 10 mM dNTP mix.
7. Superscript™ II reverse transcriptase with 5× reaction buffer and 0.1 mM dithiothreitol (DTT) (Life Technologies).
8. RNaseOUT™ (Life Technologies).
9. Sterile deionized water (dH_2O) nuclease-free.
10. 200 μL PCR tubes and a PCR tube rack.

2.1.3 cDNA Stock Dilution

1. Accurately calibrated pipettes with appropriated sterile pipette tips.
2. 8-well multichannel pipettes.
3. cDNA samples.
4. Sterile dH_2O nuclease-free.
5. 8-strip 200 μL PCR tubes with lids attached.

2.1.4 Generation of the Copy Number Standard PCR products

1. Accurately calibrated pipettes with appropriated sterile pipette tips.
2. Real-Time qPCR system—we use a LightCycler® 480 Real-Time PCR System (Roche Applied Science, Madison, WI, US) and Rotorgene RG600 (QIAGEN GmbH, Hilden, Germany) systems, though any modern qPCR thermal cycler that can export raw Ct value data would be acceptable.
3. Fluorometer or spectrophotometer for calculating DNA concentrations in a variety of solutions (H_2O, Tris EDTA-based buffer, etc.)—for example a Qubit® 2.0 Fluorometer (Life Technologies).
4. Diluted cDNA sample(s) known from plant(s) known to express the transgene.
5. Control Gene oligonucleotide primers (*see* Table 1) diluted to a working concentration of 4 μM.
6. Gene of interest (GOI) oligonucleotide primers (*see* Table 2) diluted to a working concentration of 4 μM.
7. 2× qPCR mastermix containing SYBR green, *taq* polymerase, buffer, and dNTPs—for example iQ™ SYBR® Green Supermix (Bio-Rad, Hercules, CA, USA).

Table 1
Control gene oligonucleotide primers for transgene expression determination in selected species

Species	Target	Forward primer nucleotide sequence (5′–3′)	Reverse primer nucleotide sequence (5′–3′)	Amplicon size (bp)
Arabidopsis	Glyceraldehyde-3-phosphate dehydrogenase	TGGTTGATCTCGTTGTGCAGGTCTC	GTCAGCCAAGTCAACAACTCTCTG	262
Arabidopsis	Cyclophilin	TGGCGAACGCTGGTCCTAATACA	CAAAAACTCCTCTGCCCCAATCAA	223
Arabidopsis	Actin	GAGTTCTTCACGCGATACCTCCA	GACCACCTTTATTAACCCCATTTACCA	180
Arabidopsis	Alpha tubulin	ATGTGGGTCAGGGTATGGAA	CCGACAACCTTCTTAGTCTCCTCT	143
Arabidopsis	Elongation factor alpha	AAGAAACTCACAATCATGATGACAA	AAGAAGAGACTTTAGCGAAACACAA	206
Arabidopsis	*Arabidopsis thaliana* pentatricopeptide repeat-containing protein (AT5G55840) mRNA	TGCCGAGGTGTTGAAGTTGAC	CAAGCGAGATTCCGAGAGAGG	124
Arabidopsis	*Arabidopsis thaliana* methyl-CpG-binding domain protein 4 (AT3G07930) mRNA	GGGTGTGATTCTGATAGTGTTTG	AGTTGATGCCTGGAAGTATGG	93
Barley	Heat Shock protein 70	CGACCAGGGCAACCGCACCAC	ACGGTGTTGATGGGGTTCATG	108
Barley	Glyceraldehyde-3-phosphate dehydrogenase	GTGAGGCTGGTGCTGATTACG	TGGTGCAGCTAGCATTTGAGAC	198
Barley	Cyclophilin	CCTGTCGTGTCGTCGGTCTAAA	ACGCAGATCCAGCAGCCTAAAG	122
Barley	Alpha tubulin	AGTGTCCTGTCCACCCACTC	AGCATGAAGTGGATCCTTGG	248
Rice	Actin	GAAGATCACTGCCTTGCTCC	CGATAACAGCTCCTCTTGGC	249
Rice	18S ribosomal protein	ATGATAACTCGACGGATCGC	CTTGGATGTGGTAGCCGTTT	169

Rice	Alpha tubulin	TACCGTGCCCTTACTGTTCC	CGGTGGAATGTCACAGACAC	234
Rice	Glyceraldehyde-3-phosphate dehydrogenase	GGGCTGCTAGCTTCAACATC	TTGATTGCAGCCTTGATCTG	190
Bread wheat	Actin	GACAATGGAACCGGAATGGTC	GTGTGATGCCAGATTTTCTCCAT	236
Bread wheat	Cyclophilin	CAAGCCGCTGCACTACAAGG	AGGGGACGGTGCAGATGAA	227
Bread wheat	Glyceraldehyde-3-phosphate dehydrogenase	TTCAACATCATTCCAAGCAGCA	CGTAACCCAAAATGCCCTTG	220
Bread wheat	Elongation factor alpha	CAGATTGGCAACGGCTACG	CGGACAGCAAAACGACCAAG	227
Maize	Elongation factor alpha	GCCGCCAAGAAGAAATGATGC	CGCCAAAAGGAGAAATACAAG	220
Maize	Glyceraldehyde-3-phosphate dehydrogenase	GACAGCAGGTCGAGCATCTTC	GTCGACGACGCGGTTGCTGTA	114
Maize	Actin	CCAATTCCTGAAGATGAGTCT	TGGTAGCCAACCAAAAACAGT	156
Maize	Tubulin	GAGGACGGCGACGAGGGTGAC	CAAAGCGGGGGAATAAAGTCT	186

Table 2
Selected GOI oligonucleotide primers for transgene expression determination

Target	Forward primer nucleotide sequence (5′–3′)	Reverse primer nucleotide sequence (5′–3′)	Amplicon size (bp)
Nopaline synthase (NOS) terminator	GGAGCTCGAATTTCCCCGAT	ATCATCGCAAGACCGGCAA	82
Green fluorescent protein (GFP)	GGCTCGTATGTTGTGTGGA ATTG	GGATGGTGGAGTGATGAA GGAA	187

8. PCR product purification kit—for example MinElute PCR Purification Kit (Qiagen).
9. Fluorometric DNA quantification kit—for example Qubit® dsDNA BR Assay Kit (Life Technologies), for use with the Qubit® 2.0 Fluorometer.
10. Sterile dH_2O.
11. Appropriate plasticware for the Real-Time PCR system used. This includes PCR tubes/rotors and lids/optical sealing film.
12. Sterile 2 mL microtubes for preparation of individual mastermixes.

2.1.5 Preparation of a Dilution Series for the Generation of a Standard Curve

1. Accurately calibrated pipettes with appropriated sterile pipette tips.
2. Stock copy number standard solutions for each primer set (control genes and GOI).
3. Sterile dH_2O.
4. Sterile 200 μL PCR tubes.

2.1.6 Generation of qPCR Data from Unknown cDNA Samples

1. Accurately calibrated pipettes with appropriate sterile pipette tips.
2. 8-channel and 12-channel multichannel pipettes for loading 96- and 384-well qPCR plates.
3. Real-Time qPCR system—for example a LightCycler® 480 Real-Time PCR System (Roche) and Rotorgene RG600 (Qiagen) systems, though any modern qPCR thermal cycler that can export raw Ct value data would be acceptable.
4. Diluted copy number standard solutions for each primer set (control genes and GOI).
5. Diluted unknown cDNA samples.
6. Control gene oligonucleotide primers (*see* Table 1) diluted to a working concentration of 4 μM.
7. GOI oligonucleotide primers (*see* Table 2) diluted to a working concentration of 4 μM.

8. 2× qPCR mastermix containing SYBR green, taq polymerase, buffer, and dNTPs—for example.
9. Appropriate plasticware for the Real-Time PCR system used. This includes PCR tubes/rotors and lids/optical sealing film.
10. Sterile 2 mL microtubes for preparation of individual mastermixes.

2.1.7 Calculation of Levels of Transgene Expression

1. Computer with Microsoft Excel software.
2. Gene expression analysis spreadsheet for Excel (refer to additional book materials at http://extras.springer.com).

2.2 Determination of Transgene Copy Number

2.2.1 Genomic DNA Extraction

1. Accurately calibrated pipettes with appropriated sterile pipette tips.
2. Alpha 1-2/LD Freeze dryer (Martin Christ GmbH, Germany).
3. Type MM 300 Retsch mill (Qiagen).
4. Appropriate sized stainless steel ball bearings (one or two per sample) for grinding tissue.
5. Water-bath or oven set at 65 °C.
6. Refrigerator set at approximately 4 °C.
7. Ice in a suitable container.
8. Liquid nitrogen in a suitable container—liquid nitrogen must be handled in a safe manner and the risks of using and transporting it must be understood and accounted for prior to use.
9. Extraction buffer: 100 mL 1 M Tris–HCl pH 7.5 (121.1 g Trizma base made up to 1 L with dH_2O, pH adjusted to 7.5); 100 mL 0.5 M EDTA pH 8.0 (186.1 g Na_2EDTA made up to 1 L with dH_2O, pH adjusted to 8.0); 125 mL 10 % SDS (100 g sodium dodecyl sulfate made up to 1 L with dH_2O); and 675 mL dH_2O to make up to 1 L (*see* **Note 1**). 6 M Ammonium acetate (stored at 4 °C).
10. Isopropanol.
11. 70 % ethanol (70 mL ethanol made up to 100 mL with dH_2O).
12. Sterile dH_2O for DNA resuspension.
13. 1.1 mL 8-strip microtubes with separate 8-strip caps in 96-well racks for sample collection.
14. Disposable paper towels.

2.2.2 Serial Dilution of Single-Copy Reference Sample DNA

1. Accurately calibrated pipettes with appropriate sterile pipette tips.
2. Fluorometer or spectrophotometer for calculating the DNA concentration in a variety of other solutions (H_2O, Tris EDTA-based buffer, etc.)—for example Qubit® 2.0 Fluorometer.

3. Genomic single copy reference (SCR) DNA.
4. Fluorometric DNA quantification kit—for example Qubit® dsDNA BR Assay Kit, for use with the Qubit® 2.0 Fluorometer.
5. Sterile dH_2O.
6. TE buffer: 10 mM Tris pH 8.0 with HCl, 1 mM EDTA.
7. 1.5 mL microfuge tubes.

2.2.3 Generation of qPCR Data from SCR and Unknown Genomic DNA Samples

1. Accurately calibrated pipettes with appropriate sterile pipette tips.
2. 8-channel and 12-channel multichannel pipettes for loading 96- and 384-well qPCR plates.
3. Real-Time qPCR system—for example, a LightCycler® 480 Real-Time PCR System (Roche) and Rotorgene RG600 (QIAGEN) systems, though any modern qPCR thermal cycler that can export raw Ct value data would be acceptable.
4. Serial-diluted SCR samples.
5. Genomic DNA from unknown samples.
6. Reference Gene oligonucleotide primers (working concentration = 10 μM) and dual-labeled TaqMan probe (working concentration = 10 μM) (*see* Table 3).
7. GOI oligonucleotide primers (working concentration = 10 μM) and dual-labeled TaqMan probe (working concentration = 10 μM) (*see* Table 4).
8. 2× qPCR mastermix containing taq polymerase, buffer, and dNTPs—for example KAPA Probe Fast qPCR Master Mix (2×) universal.
9. Use appropriate plasticware for the Real-Time PCR system. This includes PCR tubes/rotors and lids/optical sealing film.
10. Sterile 2 mL microtubes for preparation of individual mastermixes.

2.2.4 Calculation of Genomic Transgene Copy Number for Each Unknown Sample

1. Computer with Microsoft Excel software.
2. Genomic copy number analysis spreadsheet for Excel (refer to additional book materials at http://extras.springer.com).

3 Methods

3.1 Quantification of Transgene Expression

Quantification of transgene expression using Real-Time Quantitative PCR is calculated relative to that of stably expressed endogenous genes (control genes). Such a method allows for variable amounts of cDNA to be added to each PCR. To ensure the accuracy of the relative quantification method, the expression of

Table 3
Reference Gene oligonucleotide primers and TaqMan probes for transgene copy number determination in selected species

Species	Target	Designation	Nucleotide sequence (5′–3′)	5′ label	3′ label	Reference
Bread wheat/ barley	Puroindoline-b	Forward primer	ATTTTCCAGTCACCTGGCCC	–	–	[9]
Bread wheat/ barley	Puroindoline-b	Reverse primer	TGCTATCTGGCTCAGCTGC	–	–	[9]
Bread wheat/ barley	Puroindoline-b	TaqMan Probe	ATGGTGGAAGGGCGGCTGTGA	FAM	BHQ	[9]
Rice	Sucrose phosphate synthase gene	Forward primer	TTGCGCCTGAACGGATAT	–	–	[10]
Rice	Sucrose phosphate synthase gene	Reverse primer	CGGTTGATCTTTTCGGGATG	–	–	[10]
Rice	Sucrose phosphate synthase gene	TaqMan Probe	GACGCACGGACGGCTCGGA	FAM	BHQ	[10]

Table 4
GOI oligonucleotide primers and TaqMan probes for transgene copy number determination

Target	Designation	Nucleotide sequence (5′–3′)	5′ label	3′ label
Nopaline synthase (NOS) terminator	Forward primer	CTTAAGATTGAATCCTGTTGCCGGTC	–	–
Nopaline synthase (NOS) terminator	Reverse primer	CGAATTCAGTAACATAGATGACACCGC	–	–
Nopaline synthase (NOS) terminator	TaqMan Probe	AGCGCGCAAACTAGGATAAA	FAM	BHQ
Hygromycin resistance gene (Hyg)	Forward primer	CGCTCGTCTGGCTAAGATCG	–	–
Hygromycin resistance gene (Hyg)	Reverse primer	AGGGTGTCACGTTGCAAGAC	–	–
Hygromycin resistance gene (Hyg)	TaqMan Probe	TGCCTGAAACCGAACTGCCCGCTG	FAM	BHQ

four control genes is quantified, with the three that show the highest stability of expression across all cDNA samples selected via an inbuilt algorithm developed by Vandesompele and colleagues [1]. Initially, PCR products are generated for the transgene and each of the four control genes. A cDNA template or mixture of templates is used that is known to contain the transgene and endogenous control gene sequences to generate these products. At this stage the melt or dissociation curve is checked for an indication that a single discrete PCR product is being generated. These PCR products are purified for use as absolute cDNA copy-number standards in future reactions. Sequencing of these purified products also serves to confirm that the product is composed of a single unique sequence and not a mixture of amplified products. Following confirmation of purity, the concentrations of the PCR products are accurately quantified and a dilution series of known PCR product copy number prepared. qPCRs are performed for each primer set with the standard curve generated from PCR product copy-number standard series dilution used to calculate the absolute number of target cDNA copies present in each reaction. The number of copies of the transgene cDNA in the PCR is normalized against the most stable control genes via an Excel worksheet. An explanation of the terminology used for quantifying transgene expression is provided (*see* **Note 2**).

3.1.1 Total RNA Extraction

Similar to genomic DNA extraction, a range of methods and protocols are available to extract high-quality total RNA or purified messenger RNA from plant tissues. The method described herein is adapted from the protocol described in the TRIzol Reagent Manual (Life Technologies) and produces good quantities of very-high-quality RNA.

1. Collect one piece of leaf tissue (approximately 5–10 mm by 20–30 mm) or a small amount of root tissue, place into a collection microtube and carefully snap-freeze in liquid nitrogen (*see* **Note 3**).
2. In a fume cupboard, add 500 μL TRIzol reagent to the microtube, ensuring that the tissue does not thaw (*see* **Note 4**).
3. Grind the tissue against the side of the microtube with the plastic pestle then add a further 500 μL TRIzol reagent.
4. Incubate at room temperature for 5 min.
5. Add 200 μL chloroform to each microtube and ensure that the caps are tightly sealed and shake vigorously for 15 s (*see* **Note 5**).
6. Centrifuge at 12,000 × *g* at 4 °C for 15 min.
7. Carefully transfer the top aqueous phase to a sterile 1.5 mL microtube with a pipette, ensuring that none of the interphase and lower phase material is present. Discard the interphase and lower phase using correct chemical waste procedures.

8. To precipitate the RNA, Add 500 μL isopropanol to this solution, cap the tube and invert to mix.
9. Incubate at room temperature for 10 min.
10. Pellet the precipitated RNA by centrifuging at 12,000 × *g* at 4 °C for 10 min. Align the tubes in the centrifuge so their hinges are facing outward, allowing easy location of the RNA pellet.
11. Carefully remove the supernatant with a pipette, ensuring the RNA pellet remains in the tube (*see* **Note 6**).
12. Add 1 mL 75 % ethanol and centrifuge at 7,500 × *g* at 4 °C for 5 min.
13. Remove as much of the supernatant as possible with a pipette, while retaining the RNA pellet.
14. Leave the pellet to air-dry for 5 min.
15. Resuspend the pellet in 30 μL sterile RNase-free dH_2O (*see* **Note 7**).
16. Once fully dissolved, quantify the RNA concentration and verify quality using a spectrophotometer. The concentration of the RNA should be above 150 ng/μL and the 260/280 ratio for each sample should be between 1.8 and 2.1.

3.1.2 cDNA Synthesis

cDNA reverse-transcribed from RNA extracted in the previous protocol is the template used to indirectly quantify transgene expression. The protocol described herein uses an oligo-dT primer that anneals to the poly-A tail of mRNAs and is the initiation point for reverse-transcription. Random hexamers which prime reverse-transcription at random points across the mRNA may also be used alone or mixed with the oligo-dT primer at the same concentration. This protocol is adapted from the Superscript™ II reverse transcriptase product manual (Life Technologies).

1. For DNase I treatment, in a PCR tube add 1 μg RNA, 0.5 μL DNase I, 0.8 μL 10× reaction buffer and make up to 8 μL total volume with nuclease-free dH_2O (*see* **Note 8**).
2. Place the tubes in a thermal cycler and incubate at 37 °C for 20 min, heat-inactivate at 65 °C for 10 min, then hold at 4 °C until ready to proceed to the next step.
3. With the tubes in a PCR rack on ice, add 1 μL 50 μM oligo $(dT)_{18}$ primer, 1 μL 10 mM dNTP mix and 3 μL nuclease-free dH_2O (*see* **Note 9**).
4. Place the tubes in a thermal cycler and incubate at 65 °C for 5 min followed by holding at 4 °C until ready to proceed to the next step.
5. With the tubes in a PCR rack on ice, add 4 μL 5× Superscript™ II reverse transcriptase reaction buffer, 1 μL 0.1 mM DTT,

1 μL RNaseOUT and 0.25 μL Superscript™ II. A mastermix containing these reagents can be preprepared.

6. Place the PCR tubes in a thermal cycler and incubate at 25 °C for 5min, then 50 °C for 45 min followed by holding at 4 °C.
7. cDNA should be carefully stored at –20 °C until ready for use.

3.1.3 cDNA Dilution Method

Prior to use in quantitative PCRs, the generated cDNA should be diluted in dH_2O. Generally a 1 in 20 dilution is ideal, but where the supply of RNA is limited and the transgene to be analyzed is likely to be highly expressed, a 1 in 100 dilution can also be used. 8-strip PCR tubes and multichannel pipettes can be used to substantially increase throughput.

1. Transfer all cDNA samples to 8-strip PCR tubes in the order they will be analyzed in the qPCRs (*see* **Note 10**).
2. For 1:20 dilution, add 95 μL nuclease-free dH_2O to the appropriate number of PCR tubes.
3. Transfer to these tubes 5 μL of each cDNA stock solution using a multichannel pipette and mix well.
4. Stock and diluted cDNA can be stored at –20 °C (*see* **Note 11**).

3.1.4 Generation of the Copy Number Standard PCR Products

Copy Number Standard PCR products for each control gene primer set and the GOI primer set are generated in a qPCR reaction and checked to ascertain that single discrete PCR products have been generated. These products are purified, sequenced and quantified. Following series dilution to a set number of copies per reaction, the copy number standard PCR products act as template for the generation of a standard curve in each qPCR run. Accuracy and good quality control during this stage is essential for obtaining reliable results from the transgene expression analysis itself.

1. Thaw cDNA samples to be used for Copy Number Standard PCR product generation.
2. Mix cDNAs if required to ensure GOI product generation (*see* **Note 12**).
3. Set up four 20 μL PCRs for each control gene primer set and the GOI primer set as follows:

Forward primer (4 μM)	2.4 μL
Reverse primer (4 μM)	2.4 μL
iQ™ SYBR® Green Supermix	10 μL
cDNA sample	2 μL
dH_2O	3.2 μL
Total	20 μL

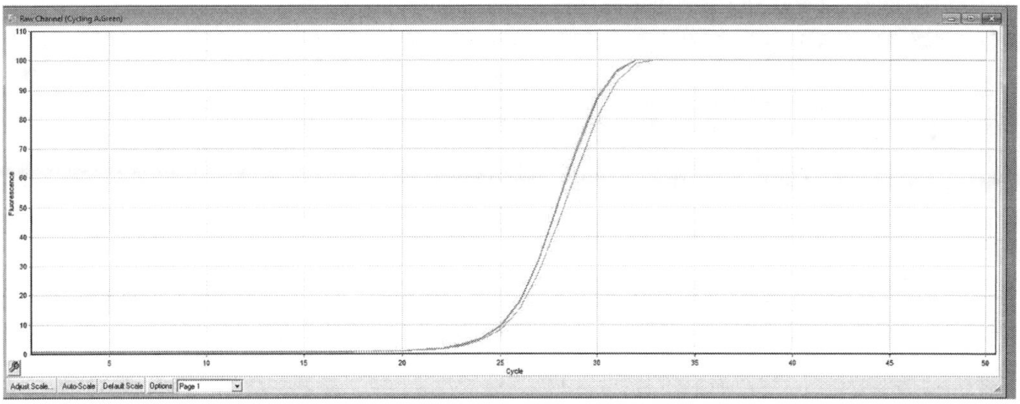

Fig. 1 A plot of raw fluorescence vs. cycle number should yield an S-shaped curve for each successful PCR. In this example, four technical replicate PCRs have been performed, with little variation in generated curves evident

4. Program the Real-Time PCR system software as follows:

95 °C	3 min
Then 40 cycles of:	
95 °C	1 s
60 °C	1 s
72 °C	30 s—set the fluorescent acquiring point here.
Melt curve:	
Increase temperature from 60 to 95 °C with fluorescence readings acquired at 0.5 °C increments.	

5. Perform the qPCR run and enter the names, details and positions of all samples into the software.
6. Use the Real-Time PCR system's software to confirm that a PCR product has been generated for each reaction from each primer set. A plot of raw fluorescence vs. cycle number should yield an S-shaped curve for each successful PCR (*see* Fig. 1).
7. Use the Real-Time PCR system's software to plot a melt curve for each PCR. Visually analyze these plots to confirm that each trace has a single peak, indicating each PCR product is likely comprised of a single nucleotide sequence (*see* Fig. 2).
8. Combine each PCR product successfully generated from the same primer set and column purify following the manufacturer's instructions. Elute the purified product in 20 μL sterile H_2O (*see* **Note 13**).
9. Accurately quantify the DNA concentration of the samples. An example is the use of Life Technologies Qubit® fluorometer and the Qubit® dsDNA BR Assay Kit according to the manufacturer's instructions to ensure accuracy.

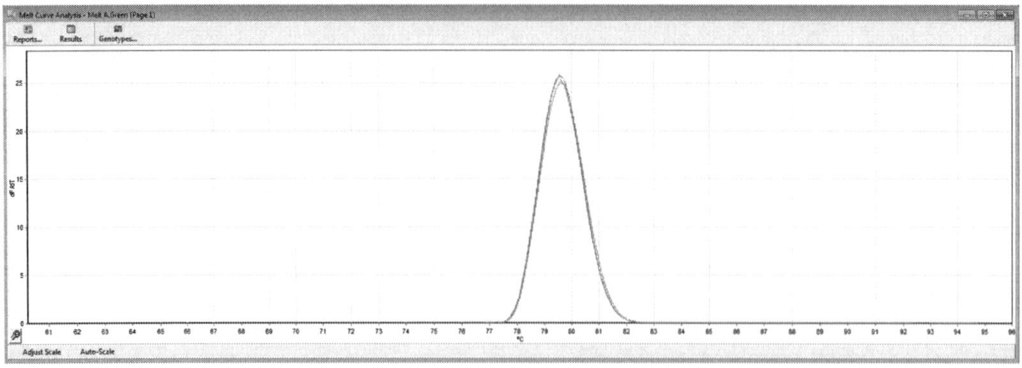

Fig. 2 Visual analysis of melt curves generated via qPCR software following the completion of the qPCRs should confirm that each trace has a single discrete peak. In addition, technical replicate PCRs should display matching melt curves. This indicates the specificity of the oligonucleotide primers used, as each PCR product is likely comprised of a single nucleotide sequence

10. Sequence the PCR products using standard protocols for your provider. Use each PCR primer and sequence the product from both termini (*see* **Note 14**).
11. Once satisfied with the purity of the PCR product, calculate the PCR product's molecular weight based on its sequence. Convert the PCR product's determined concentration in solution (ng/μL) to copies/μL (calculation). For 1×10^9 copies/μL, dilute 20/(([Concentration (ng/μL)]/1,000,000,000)/([Size (bp)] × 660) × 602,200,000,000,000,000,000,000) × 10,000,000,000 μL of the product in TE buffer to a final of 200 μL (*see* **Note 15**).
12. Prepare a stock copy standard solution of each PCR product by making a dilution in TE buffer to 1×10^9 copies/μL. A total volume of 200 μL is sufficient for a large number of experiments. Stocks should be stored at −20 °C.

3.1.5 Preparation of a Dilution Series for the Generation of a Standard Curve

Copy standard dilution series can prepared fresh or undergo a limited number of freeze–thaw cycles.

1. Make a 1×10^7 copies/μL PCR product solution by adding 2 μL of the 1×10^9 copies/μL stock solution to 198 μL dH_2O. This should be completed for all control genes and the GOI.
2. Perform six tenfold serial dilutions (1×10^7 copies/μL 1×10^2 copies/μL) by sequentially removing 5 μL of one copy standard and adding it to 45 μL dH_2O. For example, removed 5 μL from the a 1×10^7 copies/μL solution and add it to 45 μL dH_2O to make a 1×10^6 copies/μL solution, then take 5 μL of this solution and add to 45 μL dH_2O to make a 1×10^5 copies/μL solution and so forth.

3.1.6 Generation of qPCR Data from Unknown cDNA Samples

For each qPCR run, samples loaded should include the standard curve dilution series, a no-template (NTC) water (sterile dH_2O) control, and the unknown samples. All reactions should be performed in triplicate (i.e., three qPCRs per sample). Depending on the instrument used and the number of unknown samples, multiple primer sets can be used in a single run or unknown samples can be analyzed in multiple runs using the same primer set. In this case, a standard curve and NTC should be included in every run. Once PCRs have been completed with the control gene primers, they do not need to be repeated unless new cDNA samples are generated. A mastermix should always be produced for the standard curve and unknown samples to ensure the most even conditions and accuracy of the final results. For each PCR run there are 18 standard curve (six copy number standard dilutions multiplied by three replicates) and three NTC reactions (one no-template control reaction multiplied by three replicates). All other available wells can be comprised of triple replicated reactions loaded with unknown samples.

1. Set up a master mix of sufficient size for each of the four control gene primers sets and the transgene primer set (*see* **Note 16**). Each single PCR within the mastermix should comprise of:

Forward primer (4 μM)	1.2 μL
Reverse primer (4 μM)	1.2 μL
iQ™ SYBR® Green Supermix	5 μL
cDNA sample	2 μL
dH_2O	0.6 μL
Total	10 μL

2. Program the Real-Time PCR system software as follows:

95 °C	2 min

   (a) Then 50 cycles of:

95 °C	1 s
60 °C	1 s
72 °C	25 s—set the fluorescent acquiring point here.

   (b) Melt curve:

   Increase temperature from 70 to 99 °C with fluorescence readings acquired at 1 °C increments.

3. The identities of the samples in each well should be entered into the Real-Time PCR system's user interface. The standard curve samples should be indicated as such and the absolute number of copies in each reaction entered (*see* **Note 17**).

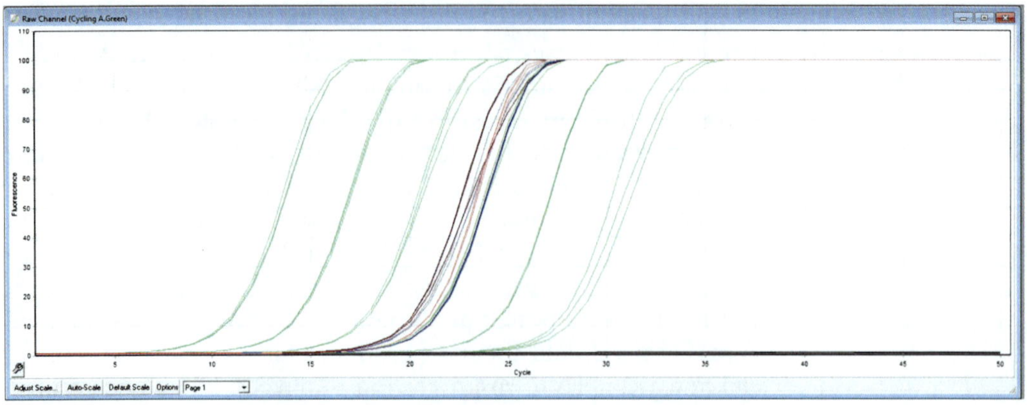

Fig. 3 A plot of raw fluorescence vs. cycle number demonstrates that the curves of the unknown samples (multicolored) fall within the range of the standard curve (*green*), allowing for accurate determination of gene expression

4. At the completion of the qPCR run, use the Real-Time PCR system's user interface according to the manufacturer's instructions to calculate a Ct value and absolute copy number for each sample. The raw fluorescence readings for each sample should fall within the range of the standard curve readings (*see* Fig. 3).
5. Visually analyze the melt curve plot for each sample, ensuring that a single peak displayed is visually similar to the peak present when the copy standard PCR product was generated earlier (*see* **Note 18**).
6. If satisfied with the melt curve plots, use the Real-Time PCR system's user interface to export the quantification data in Microsoft Excel format (.xlxs) if capable or plain text format (.txt).

3.1.7 Calculation of Levels of Transgene Expression

To calculate a value for copies of transgene cDNA per sample which can be compared across samples, the data produced using the transgene specific primers is normalized against the geometric mean of the three most stable reference genes, with the least stable excluded from the calculation. The method used is modified from Vandesompele et al. [1] and further described by Burton et al. [2]. Once the raw copy number data generated by the Real-Time PCR system's user interface is entered for the transgene and the four reference genes in the appropriate positions in the transgene expression calculation Excel workbook, the three gene normalization algorithm is automatically conducted and globally normalized transgene cDNA copy number results calculated for each individual cDNA sample. The normalized absolute copy number for any one sample can then be compared with that of any other sample, as the initial amount of cDNA loaded into each reaction no longer impacts on the result. When biological replicates have been used, the geometric mean can be calculated from the normalized absolute copy number of each replicate along with the standard error (Standard Deviation of the replicates/√no. of replicates), which can then be graphed using the Excel software. Due to the varying possible setup of the replicates within the experiment, these steps must be carried out manually.

1. Cut and Paste the absolute copy number data into the appropriate tab of the Excel spread sheet. Ensure that each set of control gene data is pasted into a control gene tab and the GOI data in the GOI tab (*see* **Note 19**).
2. In the Normalized Data Tab, fill in the names of each cDNA sample in the appropriate position (*see* **Note 20**).
3. Once completed, the normalized absolute transgene copy number and standard deviation of the technical replicates is displayed.
4. If biological or other replication has been used, the geometric mean of the replicates can be calculated manually by using Excel's GEOMEAN function and selecting the normalized transgene expression data for each set of biological replicates. The standard error of the mean (SE) can be calculated using the standard deviation (STDEV) function divided by the square root of the number of biological replicates.
5. These data can be manually presented in a bar graph with the error bars presented as ±1 SE.

3.2 Determination of Transgene Genomic Copy Number

Stable transformation of a plant involves the integration of one or more copies of a transgene into a plant's genome. Such insertion events result in a primary transformant that is hemizygous for the transgene at a particular locus (i.e., the transgene has no allelic counterpart). In subsequent generations, segregation takes place, affecting the number of copies of the transgene in the progeny. The number of copies integrated into the plant's genome at different loci during transformation is often related to the transformation methodology used; particle bombardment often results in many integration events (up to 20 or more) whilst Agrobacterium-mediated transformation generally results in fewer than five independent integrations [3, 4]. To further complicate matters, the transgene itself may undergo rearrangement along with partial insertions, deletions and duplications, particularly the in the case of particle bombardment [5]. Sequencing of integration loci and their flanking regions is the only method that can provide a comprehensive overview of transgene copy number and the degree to which the transgene remains intact. Due to the very high cost and difficulty of the method along with a very low throughput, this procedure is rarely used to analyze transgene integration on a large scale. Molecular methods such as qPCR and Southern hybridization can however, provide an estimate of the number of integrated copies of a particular target portion of the transgene. Southern-based estimation of copy number is often highly accurate, except in the case of concatenation of the transgene, where multiple transgene copies integrate directly adjacent to each other and may appear as a single copy [6]. Real-time PCR-based methods are often simpler and more rapid to perform, less labor intensive and are not confounded by concatenation of the transgene. However, it should be noted the accuracy of such methods can be adversely affected by the inefficient primer and probe performance due to design deficiencies, the sequence-related secondary structure of the transgene locus and the flanking genomic region.

The protocol described here is based on the $2^{-\Delta\Delta Ct}$ method and is adapted for amplification efficiency based on Yuan et al. [7]. Genomic DNA is the substrate for TaqMan PCRs performed using primers and probes specific to endogenous reference genes for each plant species along with primers and probes specific to the transgene itself. An explanation of the terminology used for determining transgene copy number is provided (*see* **Note 2**).

3.2.1 Genomic DNA Extraction

There are a multitude of protocols and kits available for extracting genomic DNA of sufficient quality for qPCR from plant tissue. Many of these methods can be used providing the final DNA product is suspended in a solution that does not inhibit PCR (e.g., dH_2O), PCR inhibitors from the extraction process itself are removed and the resultant genomic DNA remains intact. Here we describe a protocol that is inexpensive, simple to perform and provides DNA of sufficient quality for transgene copy number analysis [8]. As qPCR is very sensitive, large quantities of genomic DNA are not required. Overly concentrated or overly dilute DNA samples can affect the accuracy of the final results.

1. Collect two pieces of leaf tissue from a 2 to 3 week old seedlings (approximately 45 mm by 10 mm) and place into a collection microtube resting in its plate on ice.
2. Carefully snap-freeze the leaf tissue by placing plate containing the microtubes in liquid nitrogen then freeze-dry overnight in the Freeze Dryer following the manufacturer's instructions (*see* **Note 21**).
3. Add one stainless steel ball bearing to each microtube and grind the tissue in the Retsch mill for 2.5 min at a frequency of 25 revolutions per second. Check the quality of grinding and, if necessary, adjust the number of ball-bearings.
4. Grind for a further 2.5 min then carefully remove the ball-bearings (*see* **Note 22**).
5. Add 600 μL of extraction buffer to each microtube and seal with caps. Shake thoroughly. Remove the bottom of the rack and incubate the plate at 65 °C for 30 min.

6. Cool the plates at 4 °C in a refrigerator for approximately 15 min.
7. Add 300 μL 6 M ammonium acetate chilled to 4 °C. Shake vigorously to mix the ammonium acetate and incubate for 15 min at 4 °C in a refrigerator.
8. Centrifuge the plate for 15 min at 3,400×*g* to collect the precipitated proteins and plant tissue.
9. Recover 600 μL of the supernatant and transfer to new collection microtubes containing 360 μL isopropanol in each well. Cap and mix thoroughly.
10. Allow the DNA to precipitate for 5 min at room temperature.
11. Centrifuge the samples for 15 min at 3,400×*g* to pellet the DNA. Decant the supernatant and allow the remaining fluid to drain by inverting the tubes onto a paper towel for a maximum of 1 min, ensuring the retention of the pellet.
12. Wash the pellet in 400 μL 70 % ethanol by dislodging the pellet from the bottom of the microtube (*see* **Note 23**).
13. Centrifuge the rack for 15 min at 3,400×*g* and again discard the supernatant.
14. Resuspend the pellet in 400 μL of sterile dH_2O and incubate 30 min at 37 °C (incubation is optional).
15. Leave the DNA to dissolve overnight at 4 °C.

3.2.2 Serial dilution of Single-Copy Reference Sample DNA

DNA from the same species as the plant to be analyzed is used as a single-copy reference sample (SCR). Ideally, this plant should contain a single copy of the same transgene sequence as that contained in the unknown sample. Zygosity should be taken into account as, unlike Southern analysis, an SCR homozygous for one transgene integration (i.e., a T1 or later generation plant) will be quantified by qPCR as two copies rather than one.

An SCR has been selected for each plant species to control for the quantity of DNA loaded into each reaction. The copy number of this gene is stable in individual plants and provides a reference ΔCt value with which to compare the qPCR data from the transgene target. Analysis of qPCR data from serially diluted SCR samples also allows for determination of the efficiency of the PCR for both the reference gene and transgene primer/probe sets, which is later used to improve the accuracy of the final copy-number algorithm.

1. Quantify the concentration of the SCR genomic DNA sample according to the quantification equipment manufacturer's instructions.
2. Take an aliquot of the SCR genomic DNA sample and dilute using sterile dH_2O to a concentration of×ng/μL in a final volume of 100 μL.
3. Prepare three sterile 1.5 mL microtubes with 75 μL sterile dH_2O.

4. Carry out the serial dilutions by taking 25 μL of the 100 ng/μL sample and add to the first tube with 75 μL of sterile dH_2O and mix well to make a 25 ng/μL solution.
5. Take a 25 μL aliquot of the 25 ng/μL solution and add to the next sterile tube containing 25 μL dH_2O and mix well to make a 6.25 ng/μL solution.
6. Take a 25 μL aliquot of the 6.25 ng/μL solution and add to the next sterile tube containing 25 μL dH_2O and mix well to make a 1.56 ng/μL solution.

3.2.3 Generation of qPCR Data from SCR and Unknown Genomic DNA Samples

Setting up and performing the qPCRs is straightforward. qPCRs are performed in triplicate on serial-diluted SCR genomic DNAs, a non-template control (sterile dH_2O) and the genomic DNA from unknown samples using the primer/probe sets for the reference gene and the transgene.

This approach uses a TaqMan probe specific to the target sequences in addition to specific oligonucleotide primers. This technology is advantageous, as large quantities of loaded genomic DNA do not result in a reduction in dynamic range, which is often evident when using minor groove-binding dyes such as SYBR green. An additional level of target specificity is also conferred. As a TaqMan approach is used, no melt curve analysis is performed at the completion of the thermal cycling.

1. Dilute primers and TaqMan probes in sterile dH_2O to a working stock concentration of 10 μM (primers) and 10 μM (probes) (*see* **Note 24**).
2. Calculate the number of reactions to be performed for each primer/probe set. This equates to (number of SCR dilution series DNA samples) a multiple of three plus (number of unknown samples) a multiple of three plus three (no-template control).
3. Calculate the amount of reagents (not including DNA) required for a mastermix (*see* **Note 25**). A single reaction consists of:

Forward Primer (10 μM)	0.2 μL
Reverse Primer (10 μM)	0.2 μL
Probe (10 μM)	0.1 μL
Kapa Probe Fast Mix	5 μL
Sterile dH_2O	2.5 μL
DNA sample	2 μL
Total	10 μL

4. Pipette 8 μL of master mix into each PCR well.

5. Pipette 2 μL of DNA sample into the PCR well, carefully ensuring the sample is located in the correct well and that this location is recorded (*see* **Note 26**).
6. Seal the tubes/PCR plate and briefly vortex the contents.
7. Centrifuge the tubes/plates to ensure that each reaction solution is in the bottom of each well and no air bubbles are present (*see* **Note 27**).
8. Program the thermal cycler according to the qPCR master mix manufacturer's instructions, ensuring the initial denaturing step is set at the correct temperature and timing. For the Kapa Probe Fast Mix we use the following cycling parameters:

95 °C	3 min
40 cycles of:	
95 °C	2 s
60 °C	20 s—fluorescence readings acquired at this step.

9. Place the samples in the Real-Time qPCR system and initiate the run.
10. Enter the names and locations of all samples in the run in the Real-Time qPCR system's user interface. The diluted SCR samples are assigned as standards, with the concentration added. Unknowns should be assigned as unknown and the no-template control as a negative control.
11. Following the completion of the run, use the Real-Time qPCR system's user interface to automatically calculate a threshold for each primer/probe set. This allows the thermal cycler software to calculate the Ct value for each reaction. The fluorescence readings for the unknowns should occur within the range of the standard curve (*see* Fig. 4).
12. Output the sample data containing the Ct value for each reaction in the form of a text file (.txt) or Excel file (.xls or .xslx).
13. Ensure that the Ct value for the "no-template control" is either absent (no signal) or at above 35–40 cycles (*see* **Note 28**).

3.2.4 Calculation of Genomic Transgene Copy Number for Each Unknown Sample

The calculation of genomic transgene copy number is carried out automatically by the software once the Ct data is added the appropriate section of the spreadsheet. The method used is a modified $2^{-\Delta\Delta Ct}$ method that incorporates the amplification efficiency for each primer/probe set. The amplification efficiency for each primer/probe set is calculated via the Ct values from the SCR dilution series. ΔCt is calculated for a single copy by subtracting the Ct value for the reference gene from the Ct value of the transgene for each reaction in the series (the average of a 12 reactions). For an amplification efficiency of one, a difference in ΔΔCt between the CNS equates the unknown to two and indicates two genomic copies of the transgene, a difference of two cycles equates to four

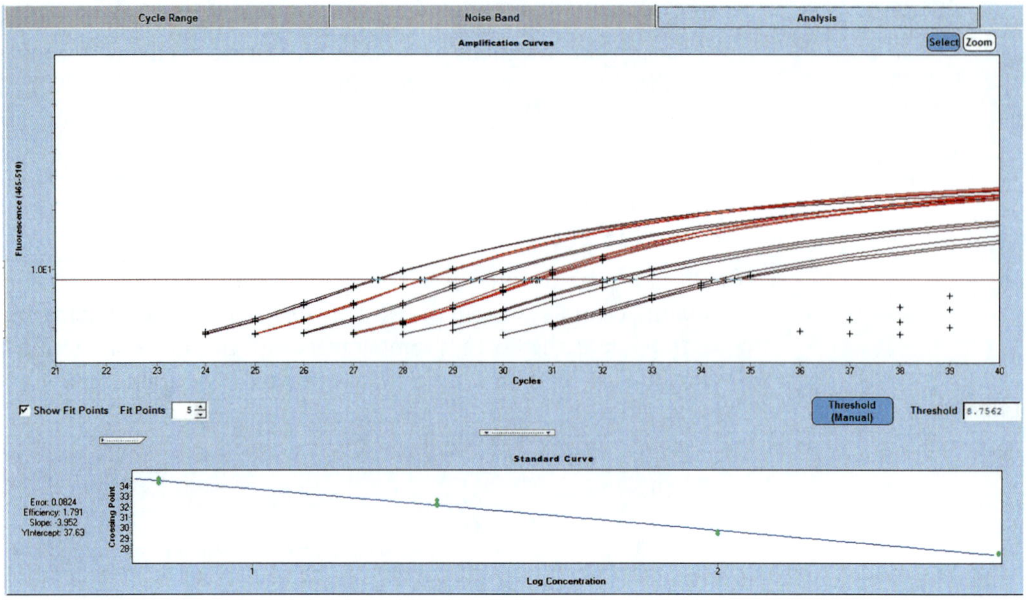

Fig. 4 A plot of adjusted curves generated by the qPCR system's user interface. All curves generated from unknown samples (*red*) fall within the range of the standard curve (*brown*). A threshold value is determined, allowing the calculation of Ct values and amplification efficiency

copies and so on. All Ct values are amplification efficiency adjusted to account for occasions where a perfect doubling of amplicons does not happen each cycle. The final result is rounded to the nearest integer. As the qPCRs can be affected by a variety of factors mentioned previously, it is important to remember that this copy number is a guide only and may not be a completely accurate reflection under all scenarios.

1. Copy and paste or enter the data for the reference gene in the reference gene tab, ensuring that the Sample names and Ct data are in the correct columns and technical replicates are in the correct rows.
2. Copy and paste or enter the data for the transgene in the transgene tab, ensuring that the Sample names and Ct data are in the correct columns and technical replicates are in the correct rows.
3. The genomic transgene copy number for each sample will be calculated automatically in the calculation tab.

4 Notes

1. This solution can be autoclaved and stored at room temperature.
2. Terminology for transgene expression analysis:
   (a) Control genes = selected endogenous reference genes.
   (b) Gene of interest (GOI) = the transgene of interest.

(c) Copy number standard = A solution containing a set number of copies of the target PCR product.

(d) Unknown sample = cDNA sample to be analyzed for GOI expression.

3. Once removed from the plant, it is essential to snap-freeze leaf tissue as rapidly as possible to prevent RNA degradation and resultant low RNA yield and quality.
4. Care should be taken opening the lid of the microtube as temperature differences can result in the lid separating with some force. Once the Trizol is added, there is no longer a requirement to keep the leaf tissue frozen.
5. When mixed properly, the solution approaches a white shade.
6. The pellet can be very difficult to see at this stage, particularly if there are few contaminants present. Maximum care must be taken to ensure that the pellet is not accidently removed.
7. Ensure that the added water is in contact with the pellet. Vortexing at this stage can help to ensure that the pellet dissolves.
8. A mastermix containing DNase I, 10× reaction buffer, and nuclease-free dH_2O can be preprepared.
9. A mastermix containing 50 μM oligo $(dT)_{18}$ primer or random hexamers, 10 mM dNTP mix, and nuclease-free dH_2O can be preprepared.
10. The correct order is important to ensure that high throughput can be achieved in the PCR loading stage by use of a multichannel pipette.
11. Handling of stock cDNA solutions should be minimized to avoid contamination and lead to inaccurate quantification results.
12. If unsure whether a particular cDNA sample will contain the target sequence for GOI product generation, aliquots of multiple diluted cDNA samples can be taken and mixed, maximizing the probability that the target sequence will be present and a PCR product generated. This mixed sample is not required for latter analysis stages.
13. A failed reaction should not be column purified. Multiple failures indicate there are problems with the primers set.
14. The amplification of the correct sequence can be confirmed by aligning the PCR product sequences with the known target sequence. In addition, the absence of mixed or unreadable sequencing results is a further indication of the purity of the PCR product and the specificity of the PCR primers to the target of interest.
15. Free online software is available for calculating PCR product molecular weight (e.g., http://www.basic.northwestern.edu/biotools/oligocalc.html). Ensure that the calculation is for double-stranded DNA.

16. An additional 10 % volume is generally acceptable for the PCR mastermix.
17. At the completion of the run, these entered details allow the software to correctly calculate a threshold cycle (Ct) value for each sample.
18. A radically different peak could indicate incorrect primers, reaction constituents, or thermal cycling conditions were used.
19. It is essential that the positioning of this input is in the correct column and rows within the Excel spreadsheet tab to allow the software to correctly calculate the mean and standard deviation of the three technical PCR replicates for each sample and for all further downstream calculations.
20. The Excel normalization algorithm uses only samples with an entered name in the normalization calculations.
21. If time constraints exist, the sealed microtubes can be stored at −80 °C overnight prior to freeze-drying.
22. At this stage racks with sealed microtubes can be stored at −20 °C prior to proceeding with the protocol.
23. Pay particular attention to ensure that the pellet does not adhere to the lid.
24. These working stocks should not be used for more than one or two freeze–thaw cycles. If in doubt, prepare new working stock solutions from the storage stocks.
25. Generally calculate to include 10 % excess in reaction numbers to ensure sufficient master mix for all reactions.
26. Accurate sample loading at this stage is essential to the precision of the final result.
27. This centrifugation step is not required if using a rotary thermal cycler as the cycler's rotor acts as a centrifuge.
28. A Ct value close to that of the DNA samples indicates contamination of the PCR reactions and a likely adverse effect on the accuracy of the results. If contamination is suspected, the PCR should be attempted again with careful technique and new dilutions of reagents.

References

1. Vandesompele J, De Preter K, Pattyn F et al (2002) Accurate normalization of real-time quantitative RT-PCR data by geometric averaging of multiple internal control genes. Genome Biol 3:1–11
2. Burton RA, Shirley NJ, King BJ et al (2004) The *CesA* gene family of barley. Trends Plant Sci 9:418–425
3. Kohli A, Twyman RM, Abranches R et al (2003) Transgene integration, organization and interaction in plants. Plant Mol Biol 52:247–258
4. Travella S, Ross SM, Harden J et al (2005) A comparison of transgenic barley lines produced by particle bombardment and *Agrobacterium-mediated* techniques. Plant Cell Rep 23: 780–789

5. Svitashev SK, Pawlowski WP, Makarevitch I et al (2002) Complex transgene locus structures implicate multiple mechanisms for plant transgene rearrangement. Plant J 32:433–445
6. Bubner B, Baldwin IT (2004) Use of real-time PCR for determining copy number and zygosity in transgenic plants. Plant Cell Rep 23:263–271
7. Yuan JS, Wang D, Stewart CN Jr (2008) Statistical methods for efficiency adjusted real-time PCR quantification. Biotechnol J 3:112–123
8. Shavrukov Y, Miyazaki J et al (2010) *HvNax3*-a locus controlling shoot sodium exclusion derived from wild barley (*Hordeum vulgare* ssp. *spontaneum*). Funct Integr Genomics 10: 277–291
9. Li Z, Hansen JL, Liu Y et al (2004) Using real-time PCR to determine transgene copy number in wheat. Plant Mol Biol Rep 22:179–188
10. Yang L, Ding J, Zhang C et al (2005) Estimating the copy number of transgenes in transformed rice by real-time quantitative PCR. Plant Cell Rep 23:759–763

Chapter 18

High-Throughput Analysis Pipeline for Achieving Simple Low-Copy Wheat and Barley Transgenics

Nataliya Kovalchuk

Abstract

Transgenic (or genetically modified—GM) plant breeding is increasingly being used as a supplementary tool to many classical plant-breeding programs. Currently the range of transgenic traits accepted for commercial use is largely restricted to herbicide and pest resistance. Given the fact that transgenics can offer an alternative and novel source of genetic variation, pre-breeding research is now increasingly exploiting this technology to tackle a greater spectrum of traits. These traits range from abiotic stress tolerance to improved product quality and nutritional characteristics. Likewise there is an increasing demand for high-throughput methodologies for transgenic plant generation, characterization and phenotyping. Selecting simple low-copy number transgenic events that are both heritable and stably expressed "in planta" is considered a prerequisite to systematic phenotyping for traits of interest. Furthermore, this assessment relies heavily on comparisons to appropriate control plants, in the case of wheat and barley transgenics this is both wild-type and null siblings. This chapter presents a general scheme on which to base selection of transgenics and respective null siblings using wheat and barley as an example. This scheme can be adapted to other similar crop species. Overall this strategy reduces the total number of plants to be genotyped and phenotyped at each generational step, and therefore resulting in significant savings in time, effort, and resources.

Key words GM plants, Gene of interest (GOI), Transformation, RNA, DNA, qRT-PCR, Homozygosity

1 Introduction

Biolistics, *Agrobacterium*-mediated and electroporation are transformation methodologies used for integrating gene sequences of interest into a host genome. These technologies have been applied to many important crop species with varying success. Until recently, cereals like wheat and barley have been recalcitrant to transformation, mainly due to low in vitro culture regeneration frequencies. However, refined protocols and genotype selection are now achieving high efficiencies of transformation routinely. Consequently there is a rapid increase in the uptake of these capabilities by research organizations and breeding programs. With increased demand there is a growing need for systematic simple

Delphine Fleury and Ryan Whitford (eds.), *Crop Breeding: Methods and Protocols*, Methods in Molecular Biology, vol. 1145, DOI 10.1007/978-1-4939-0446-4_18,

high-throughput methodologies that allow researchers and breeders to select only those transformants that contain simple, low-copy number transgene insertion events that are both heritable and express all components in *planta*. The proportion of transformation events that fit these criteria is typically low and is highly dependent on transformation methodology used. For example, biolistics often results in complex multi-copy multi-locus transgene integration patterns [1–3]. *Agrobacterium*-mediated transformation, alternatively, has a higher proportion of simple low-copy events but is variable depending upon bacterial strain used [4]. Selecting only desired transgenic events and appropriate control plants reduces the numbers of transformants to be handled, grown to maturity and ultimately propagated, saving valuable time, energy, and resources.

This chapter presents a methodological overview to aid cereal research laboratories and breeding programs in the selection of simple, low-copy wheat and barley transgenics plus their respective null sibling controls. This general methodology can be adapted to a range of crop species and be modified according to the transgene of interest.

2 Materials

2.1 Plant and Seedling Growth

1. Routinely transformable genotypes: *Hordeum vulgare* (cv Golden Promise, advanced breeding line WI4330), *Triticum aestivum* (cv Gladius).
2. Soil mix: 75 % (v/v) Coco Peat, 25 % (v/v) nursery cutting sand (sharp), 750 mg/L $CaSO_4 \cdot 2H_2O$ (gypsum) 750 mg/L $Ca(H_2PO_4)_2 \cdot H_2O$ (superphosphate), 1.9 g/L $FeSO_4$, 125 mg/L FeEDTA, 1.9 g/L $Ca(NO_3)_2$, 750 mg/L Scotts Micromax micronutrients, and 2.5 g/L Osmocote Plus slow release fertilizer (16:3:9) (Scotts Australia Pty. Ltd.). pH is adjusted to between 6.0 and 6.5 using 2 parts agricultural lime to 1 part hydrated lime.
3. Pots: 0.55 L MK12 Masrac Taglok punnet pots (90 × 80 mm), 2.8 L planting pots (6 inch diameter).
4. TPS TOP—clear plastic micropropagating cover (Thermofilm, Springvale, Australia).
5. Imaging: Digital camera, tripod, black cardboard or felt sheet.
6. Seed sterilization: 70 % Ethanol, 1 % sodium hypochlorite solution supplemented with 0.5 μL/mL Tween 20, 90 × 20 mm gamma-sterilized petri dishes, tissue paper, sterile dH_2O, forceps.
7. GA_3 solution: 1.4 mM gibberellic acid (0.5 g of Sigma G-7645 dissolved in 1 L distilled water).
8. Bags and envelopes.

2.2 Genomic DNA Extractions from Freeze-Dried Tissues

1. Alpha 1–2/LD Freeze dryer (Martin Christ GmbH, Germany).
2. Type MM 300 Retsch mill (Qiagen).
3. Extraction Buffer: 0.1 M Tris–HCl pH 7.5, 0.05 M EDTA pH 8.0, 1.25 % SDS, preheated to 65 °C.
4. 4 mm stainless steel or tungsten ball bearings.
5. 6 M Ammonium acetate stored at 4 °C.
6. 100 % Isopropanol.
7. 70 % Ethanol.
8. 95 % Ethanol.
9. 1× TE buffer: 10 mM Tris pH 8.0 with HCl, 1 mM EDTA.
10. R40 buffer: 1× TE buffer, 40 μg/mL RNase A.
11. 96-Well block containing 8 × 1.1 mL strip tubes with capping mat or strip lids.

2.3 PCR for Transgene Detection

1. MJ Research PTC100 Thermocycler (Bio-Rad).
2. 96-Well thermocycling plate, plate seal.
3. PCR reaction components: 50 mM $MgCl_2$, 10 μM dNTPs, 5 U/μL commercially available *Taq* DNA polymerase with supplied 10× PCR Buffer.
4. TE buffer: 10 mM Tris pH 8.0 with HCl, 1 mM EDTA.
5. Sterile dH_2O.
6. Transgene-specific oligonucleotides (examples):
   (a) *hpt*
   Forward primer: 5′ AGACCTGCCTGAAACCGAACTGC 3′
   Reverse primer: 5′ CACGGCCTCCAGAAGAAGATGTTG 3′
   Annealing temperature: 60 °C
   Number of cycles: 30
   Product size: 411 bp
   (b) *nptII*
   Forward primer: 5′ GAGGCTATTCGGCTATGACTG 3′
   Reverse primer: 5′ ATCGGGAGCGGCGATACCGTA 3′
   Annealing temperature: 57 °C
   Number of cycles: 30
   Product size: 679 bp
   (c) *bar*
   Forward primer: 5′ GTCTGCACCATCGTCAACC 3′
   Reverse primer: 5′ GAAGTCCAGCTGCCAGAAAC 3′
   Annealing temperature: 57 °C
   Number of cycles: 30

Product size: 444 bp

(d) CP4 *epsps*

Forward primer: 5′ TCTCGCTAGCGGTGAAACT 3′

Reverse primer: 5′ TTGAGCGGAAGCCATAGGT 3′

Annealing temperature: 52 °C

Number of cycles: 30

Product size: 400 bp

3 Methods

Biolistics and *Agrobacterium*-mediated transformation generates many unwanted transgenic events that at each generational step need to be eliminated from further genotypic and phenotypic characterization. Ideally a large number of T0 transgenic lines are reduced to just a few (~5–10) T3 homozygous low-copy, single-locus transgenic lines each representing independent transgenic events (Fig. 1). At each generational step, qPCR copy number and qRT-PCR expression analyses helps identify which individuals to retain for phenotyping, growth till maturity, and ultimately propagation (Fig. 1). The following general methodology can be presented in a structured decision-based flowchart (Fig. 2).

3.1 T0 Plantlet Growth, Tissue Sampling, and Documentation

In vitro culture, regeneration, and selection of healthy viable plantlets from embryogenic calli post-transformation varies depending upon choice of selectable marker. Selectable markers typically used for either wheat or barley transformation include *hpt*, *nptII*, *bar*, or CP4 *epsps* which encode resistance to hygromycin, kanamycin, phosphinothricin, and glyphosate, respectively. Promoters routinely used to drive selectable marker and ectopic gene overexpression in wheat and barley are rice *Actin*, maize *Ubiquitin*, and *CaMV 35S*. The use of these constitutive promoters greatly simplifies analysis for transgene expression allowing any viable tissues to be sampled and analyzed. For wheat and barley transgenics, we routinely extract DNA from young leaf blades and RNA from fully expanded mature leaf blades. This material is used for copy number and expression analysis of transgenes.

1. Transfer only those T0 regenerants (plantlets) from in vitro culture to glasshouse conditions that express resistance to the selectable trait (e.g., antibiotic or herbicide resistance) encoded by the introduced selectable marker (e.g., *hpt*, *nptII*, *bar*, or CP4 *epsps*). These T0 transformants should ideally contain 1 but less than 3 copies of the target transgene (*see* **Note 1**) as determined by qPCR. Transgene copy number determination can also be achieved by Southern analysis (*see* **Note 2**). Ideally

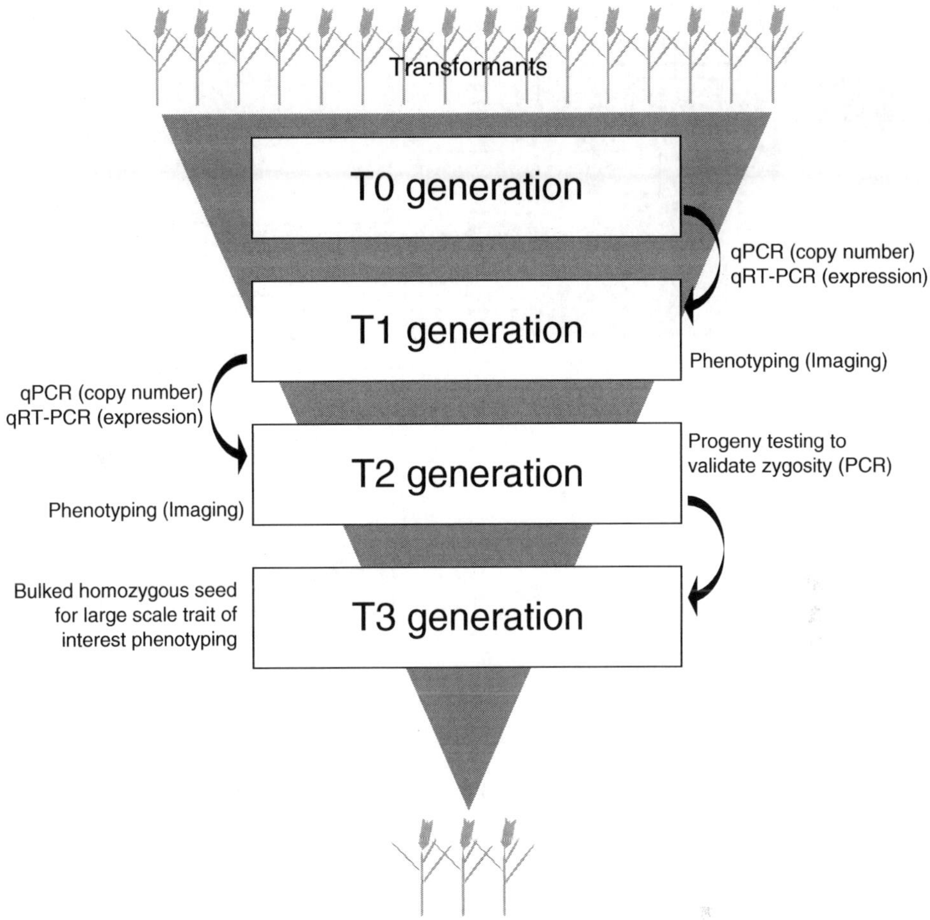

Fig. 1 Schematic overview of T0 to T3 analysis for identifying simple low-copy wheat and barley transgenics

several plantlets deemed non-transgenic by qPCR (or Southern) should be planted out as controls for subsequent molecular and phenotypic analyses (*see* **Note 3**). Place all plantlets in MK12 punnets filled with moist soil mix. Ensure that all plants are placed in a high humidity environment to allow adaptation from in vitro culture to glasshouse conditions. We use TPS TOP clear plastic micropropagating covers with adjustable vents (close) to seal potted plantlets from low-humidity glasshouse conditions.

2. Place all covered potted plantlets into a controlled glasshouse set to 20/15 °C day/night temperature with a 12 h photoperiod. If natural day length is not sufficient then supplement to achieve a photoperiod of 12 h using 400 W metal halide lamps.
3. Gradually over the period of 1-week open TPS TOP clear plastic micropropagating cover air-vents to allow transformant leaves

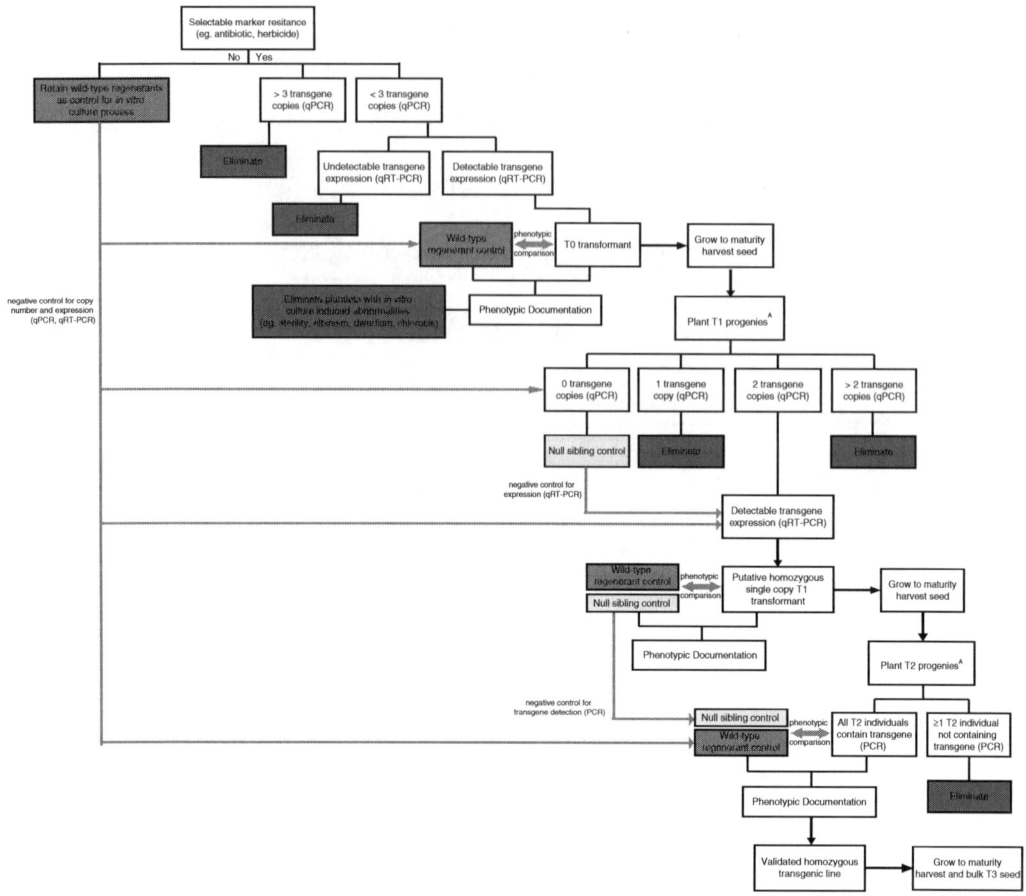

Fig. 2 Decision based flowchart for selecting simple low-copy wheat and barley transgenics. A. The numbers of T1 and T2 individuals to be analyzed should increase proportionately relative to the putative number of transgene copies/loci. Twelve individuals are routinely analyzed for wheat and barley transgenics as it facilitates 96 sample high-throughput genotyping

to "harden-off." After 1 week, TPS TOP clear plastic micro-propagating covers can be removed. Monitor and water plants routinely to avoid dehydration stress.

4. Observe plantlets relative to the non-transgenic regenerant controls for phenotypic abnormalities. At this stage, phenotypes may either be associated with the integrated transgene or be a consequence of the in vitro culture process. Depending upon the number of independent transformants generated (typically between 10 and 20) and the proportion of which exhibit the phenotypic abnormalities, it is important to decide whether to retain plantlets for further analysis. Typically a very low number (depending on construct and GOI) of transformants exhibit phenotypic abnormalities like sterility, albinism, dwarfism, or chlorosis, for species like wheat and barley. These are removed from further propagation and analysis.

5. Once plantlets have generated sufficient biomass, harvest several fully expanded leaf blades and place in 10 mL centrifuge tubes for subsequent molecular characterization at the RNA, protein, and/or metabolite levels (the level of characterization will be transgene dependent). The total amount of tissue harvested will restrict the number of downstream analyses that can be performed. Leaf tissue can be stored intact at −80 °C for several months (for RNA and protein analysis).
6. As plants approach anthesis, transgenic phenotypes can be documented relative to non-transgenic control(s). Documentation is best achieved by photographing the potted transgenic plant(s) pair-wise with its respective regenerant control(s) in front of black cardboard or felt sheet. Altered flowering time is often observed in cereal transgenics relative to controls, this should be documented to aid phenotypic assessment in subsequent generations.
7. Allow plants to grow till maturity (*see* **Note 4**).
8. Harvest mature spikes and place at 37 °C for 1 week or until material is completely dry.
9. Harvest seed, package on an individual plant basis (*see* **Note 5**) and store at 4 °C in the dark till use. Seeds can be pretreated (imbibed) with 1.4 mM GA_3 solution overnight to help remove seed dormancy and promote germination if required.

3.2 T1 Plant Growth, Tissue Sampling, and Analysis

The individual analysis of T1 plants for transgene copy number is important for determining transgene heritability, segregation and zygosity, three important criteria for selecting stable low-copy single locus insertion transgenics for phenotyping. This information collectively helps validate the early generation (T0) plant selection for simple low-copy transgenic events by qPCR and/or Southern.

1. Select 10 low-copy transgenic lines (derived from a single T0 plant) and plant 12 progeny seeds for each in 2.8 L 6 inch diameter pots containing moist soil mix. Ensure that 1 seed is planted per pot. Similarly plant wild-type and respective null controls.
2. Transfer to either glasshouse or controlled environment growth cabinet (BioChamber, BigFoot Series) set at 22–25/18 °C day/night under a 16 h photoperiod with a lighting intensity of approximately 50 μmol m^{-2} s^{-1} using a combination of Grolux fluorescent lamps and low wattage incandescent lamps.
3. Harvest 3 cm of tissue from the first true leaf blade of each plant and array in 1.1 mL, 8-strip vials in a 96-well plate format. Keep on ice for freeze-dried DNA extraction and subsequent genotyping by qPCR. Leave seedlings to continue growing.
4. If freeze-drying of harvested leaf tissue is not to be performed within 4 h samples can be stored at either −20 °C or −80 °C

(−80 °C is recommended for storage over periods more than 1 month). To ensure high-quality DNA, frozen or cooled material should be transferred directly to the freeze-dryer and left overnight.

5. Freeze-dried leaf samples can be stored at room temperature for short periods (e.g., several days) prior to grinding. Longer storage of freeze-dried samples, even at low temperature, is not recommended, as samples will rehydrate over time. This makes grinding of the tissues difficult.
6. Add one 4 mm ball bearing to each sample and seal tube. Transfer to Retsch Mill and shake vigorously for 2.5 min at 25/s. Check that all samples are ground to a very fine powder.
7. Remove ball bearings by carefully inverting plate so as to avoid sample cross-contamination. Samples can be stored at −20 °C for short term and longer term at −80 °C.
8. Add 600 μL of pre-warmed (65 °C) extraction buffer to each sample and incubate at 65 °C for 30 min.
9. Cool samples briefly at 4 °C, then add 300 μL of cool (4 °C) 6 M ammonium acetate. Mix well and incubate at 4 °C for a further 15 min.
10. Centrifuge 96-well plate(s) at 4 °C for 15 min at 4,000 rpm (>1,000 × *g*) to precipitate insoluble cellular debris. Carefully transfer 600 μL of supernatant to new 96-well plate containing 8 × 1.1 mL strip vials.
11. Add 360 μL of ice-cold isopropanol. Mix gently, allow nucleic acids to precipitate at −20 °C for 10 min to overnight. Centrifuge at 4 °C for 15 min at 4,000 rpm (>1,000 × *g*) to pellet nucleic acids. Carefully remove all supernatant ensuring not to dislodge the pellet. Allow pellet to air-dry.
12. Wash pellets with 300 μL of 70 % ethanol. Centrifuge at 4 °C for 15 min at 4,000 rpm (>1,000 × *g*). Carefully discard supernatant.
13. Wash pellets with 300 μL of 95 % ethanol. Centrifuge at 4 °C for 15 min at 4,000 rpm (>1,000 × *g*). Carefully remove all supernatant. Allow pellet to air-dry.
14. Resuspend pellet in 400 μL of either sterile dH_2O or 1/10 strength TE buffer if used directly as template for PCR or qPCR. Resuspend in 35–50 μL of R40 buffer if it is to be used for Southern analysis.
15. Genotype individual transgenic lines and controls for copy number by qPCR (*see* Chapter 21) or Southern analysis (*see* **Note 6**).
16. 3 weeks post seedling germination collect 1 fully expanded leaf blade from each transgenic plant and non-transgenic control for RNA extraction (*see* **Note 7**). Place leaf tissue in 10 mL centri-

fuge tubes then immediately snap-freeze in liquid nitrogen. Store intact frozen leaf samples for expression analysis at –80 °C.

17. Extract RNA and genotype for transgene expression (*see* Chapter 21).
18. For each transgenic line select all homozygous plants (2 transgene copies) with the desired level of transgene expression and 1 null with no detectable transgene (*see* **Notes 8** and **9**). Additional wild-type negative controls (at least 10 plants) should also be included. All other plants can be discarded from further analysis.
19. Further tissue sampling can be undertaken depending on desired additional molecular analyses (e.g., protein expression, metabolite analysis).
20. As retained plants approach anthesis, transgenic phenotypes relative to nulls and non-transgenic control(s) should be observed and documented accordingly (*see* Subheading 3.1, **step 6**).
21. Grow plants till maturity.
22. Harvest mature spikes and place at 37 °C for 1 week or until material is completely dry.
23. Harvest seed, package on an individual plant basis (*see* **Note 5**) and store at 4 °C in the dark till use. Seeds can be pretreated (imbibed) with 1.4 mM GA_3 solution overnight to help remove seed dormancy and promote germination if required.

3.3 T2 Validation for Transgene Zygosity and Integrity

The purpose of T2 segregation analysis is to validate that selected T1 individuals with 2 transgene copies as determined by qPCR are indeed homozygous for the transgene. This PCR protocol can also be used to validate whether all transgene components (selectable marker, GOI plus respective regulatory sequences) have been integrated in this homozygous line.

1. Surface-sterilize 12 T2 progenies seeds per T1 transgene expressing homozygous line by soaking in 2 mL 70 % ethanol for 5 min (*see* **Note 10**). This should be repeated for respective regenerant control(s) and the parental wild-type line. Remove ethanol and replace with 2 mL of 1 % sodium hypochlorite solution supplemented with 0.5 μL/mL Tween 20 for 15 min. Remove this solution and rinse three times for 5 min with fresh sterile dH_2O. Allow seeds to imbibe by soaking in sterile dH_2O for 3 days at 4 °C.
2. Place sterile imbibed seeds on a bed of tissue paper, one transgenic line per plate. Saturate tissue paper with sterile dH_2O. Seal petri dishes and transfer to a controlled environment growth cabinet (BioChamber, BigFoot Series) set at 22–25/18 °C day/night under a 16 h photoperiod with a lighting intensity of approximately 50 μmol m^{-2} s^{-1} using a

combination of Grolux fluorescent lamps and low wattage incandescent lamps. Leave for 4–5 days.

3. Array all seedlings in 8×12, 1.5 mL microcentrifuge tube rack(s) to facilitate sample tracking in a 96-well format (*see* Subheading 3.2, **step 2**) (*see* **Note 11**). Add approximately 0.5 mL of sterile dH_2O into each well to prevent seedling dehydration. Return to controlled environment growth cabinet for 1 day.
4. Harvest 4 cm of leaf tissue for DNA extraction (*see* Subheading 3.2, **step 2**). Remaining seedlings can stay at 4 °C for a period of 4–5 weeks, given their roots are kept moist, with little ill affect.
5. Extract DNA (*see* Subheading 3.2, **step 5–12**) and dilute to a concentration of approximately 50 ng/μL in either 1×TE buffer or sterile dH_2O. This will be the template DNA for a subsequent PCR to detect the presence of the transgene (*see* **Note 12**). For screening wheat and barley transformants we routinely use PCR to amplify the hygromycin selectable marker gene sequence (*hpt*). This protocol can be adapted to amplify any selectable marker (*see* Subheading 2.3, **step 5**) or gene of interest sequence (*see* **Note 13**).
6. Aliquot 2 μL of each template DNA into 96-well thermocycling plate(s). Prepare a PCR master mix of sufficient size for all samples to be tested. Each individual PCR reaction within the mastermix should comprise of the following components:

Forward primer (10 μM)	0.25 μL
Reverse primer (10 μM)	0.25 μL
10× PCR Buffer	1 μL
$MgCl_2$ (50 mM)	0.1 μL
dNTPs (10 μM)	1 μL
Taq DNA polymerase (5 U/μl)	0.05 μL
Sterile MilliQ H_2O	5.35 μL
TOTAL	8 μL

7. Aliquot 8 μL to each sample and seal the 96-well thermocycling plate.

   Thermocycling conditions for amplicons <500 bp (*hpt* amplicon is 411 bp): 95 °C 7 min

   Then 40 cycles of:

   95 °C 30 s

   60 °C 30 s

   72 °C 60 s

8. Resolve PCR amplicons by agarose gel electrophoresis and visualize.
9. The presence of amplicons for all 12 T2 progenies relative to the wild-type control validates that the T1 parent was homozygous for the single integrated transgene (*see* **Note 14**). If any T2 progenies do not yield an appropriately sized amplicon (e.g., 411 bp for *hpt*) this indicates that the T1 parent was hemizygous for the integrated transgene. All selected null T2 progenies should not generate amplicons (*see* **Note 15**).
10. Transfer all transgenic seedlings derived from a non-segregating family to 2.8 L 6 inch diameter pots (*see* **Note 16**) containing moist soil mix and place in a controlled environment growth cabinet or glasshouse (*see* Subheading 3.1, **step 2**).
11. Once seedlings have generated sufficient biomass they can be further sampled for additional molecular analyses (e.g., protein expression, metabolite analysis).
12. As plants approach anthesis, document (imaging) transgenic phenotypes relative to nulls and non-transgenic control(s). Because these T2 individuals have been selected for homozygosity, transgene expression should not be affected by variability in transgene dosage and plants can be reliably phenotyped relative to respective null and wild-type controls. At this stage, transgenics can be phenotyped for any developmental aberrations according to the Zadock's scale (*see* ref. [5]). This is in part achieved by documenting traits like tiller number, plant height and biomass, leaf and node number per tiller, as well as flowering time.
13. Grow plants till maturity.
14. Harvest mature spikes from all material keeping separated on a per line basis, place intact spikes at 37 °C for 1 week then weigh. Count the number of spikes derived from each plant, and then thresh and collect seed. Weight and count total number of isolated seeds per plant (*see* **Note 17**). To calculate additional indicators of plant performance related to yield, harvest, dry then weigh the remaining plant biomass. Harvest Index is the weight of the total seed per plant as a percentage of the total plant biomass (total spike weight plus total shoot weight).
15. Collect seed, package, and store on an individual line basis. At this stage all T3 seeds from homozygous T2 individuals can be bulked together and used for further testing (i.e., transgenic field trials).

3.4 Trait-Specific High-Throughput Phenotyping

High-throughput phenotypic analysis is best achieved by comparing only well-characterized T3 homozygous single-copy single-locus transgenics to respective null and wild-type controls. This can be undertaken in either large containment glasshouse facilities or alternatively under field conditions. Confined field trials are small-scale experiments undertaken to evaluate transgenic performance and trait efficacy under local environment conditions. This form of experimentation is subject to local gene technology legislations and typically performed under stringent terms and conditions that confine the experimental material either by spatial isolation or reproductive isolation [6].

Important to both containment glasshouse and confined field trails is trail design and statistical analyses; this is ultimately dependent upon the phenotyping regime for the trait of interest. However, transgenic field trails are typically across multiple locations have 3–4 replications per trial site incorporating a completely randomized block design. Such designs take into account spatial variation caused by differences in environmental factors such as soil type and moisture availability.

4 Notes

1. We recommend using just low-copy (not greater than 4) events as higher multi-copy multi-locus events can be difficult for zygosity determination.
2. Although Southern blotting techniques are traditionally used for estimation of transgene copy number and integrity of the insert, this method is difficult to apply for high-throughput screening. qPCR is used as alternative methodology because it gives quantitative and accurate copy number estimation from small quantities of isolated DNA. Only requiring small amounts of DNA facilitates screening large numbers of putative transformants from tissue culture regeneration.
3. These can be derived from either planting wild-type seedlings or identifying selectable marker escapes in the in vitro culture process. In the case of non-transgenic selectable marker escapes, they are sometimes deemed better T0 phenotyping controls as they have undergone the tissue culture process and therefore allow the exclusion of some in vitro culture induced phenotypic anomalies (i.e., somaclonal variants, vitrification).
4. Very often cereal transgenics grown to maturity directly from tissue culture generate significantly less biomass and yield less seed. This can restrict subsequent phenotypic and molecular analyses.
5. Each plant/seed packet will represent an independent transformation event. However, it should be noted that multiple regenerant plantlets can originate from the same transformed cell(s). These are termed clonal regenerants and occur at a frequency of 3–4 % for species like wheat and barley.
6. For a transgenic line genotyped to be single copy, it is assumed to be hemizygous in the T0 generation. Therefore it is expected

that T1 transformants could either be homozygous, hemizygous or null for the transgenic event. These three genotypes can be determined by qPCR to be two, one and zero copies respectively. Wild-type material and/or non-transgenic regenerants should be used as negative controls for transgene copy number determination.

7. The tissue required for transgene expression analysis is dependent upon the promoter used to drive transcription of the GOI. The methodology described here is particularly useful for transgenes that are constitutively overexpressed and expected to be expressed in the first true leaf (e.g., *CaMV 35S*, Maize *Ubiquitin*, or Rice *Actin*). Adjust tissue to be sampled according to the promoter used.

8. It is necessary to assess the proportion of two, one and zero copies of the transgene relative to the expected transgene copy number as determined by T0 qPCR (e.g., 1:2:1, for a single locus insertion). For segregation analysis on larger copy number transgenics, proportionately larger numbers of T1 progenies need to be assessed. In this case a Chi-squared test can be adopted to test for deviation in expected copy number segregation relative to that observed. If a plant containing the desired homozygous single locus insertion, plus its respective null are not identified, larger numbers of T0 progenies should be genotyped. If seed number is restricted, cross a low-copy-number T1 individual with the wild-type parent donor and genotype F2 individuals. This allows sufficient seed to be generated whilst maintaining segregation of the transgenic loci.

9. For some T1 progenies no transgene may be detected. This indicates that the transgene was not heritable and that the source plant was a likely transgenic chimera. Transgenic chimeras are the product of cellular populations recruited for shoot organogenesis in which some cells either transiently express the selectable marker, or where one selectable marker expressing cell population is able to protect an untransformed cell population. The prevalence of transgenic chimeras varies depending on selectable marker choice.

10. To facilitate surface sterilization the outer husk of barley grains can be removed to expose the underlying caryopsis. The outer husk is often a source of fungal and bacterial contamination.

11. Arraying seedlings in groups of 12 facilitates transgene segregation analysis in a high-throughput 96-well plate format. This format also simplifies downstream data processing. Twelve individuals should allow the identification of approximately 3 null segregants per single copy transgenic line. The numbers of T1 individuals analyzed should increase proportionately to the putative number of transgene copies/loci (i.e., for 3 transgenic copies/loci, analyze 36 T1 individuals).

12. Undertaking copy number qPCR is not necessary for this step.
13. Although it is preferential to use Southern analysis to confirm complete integration of the transgene into the host genome, PCR can be used as a quick viable alternative. An indication of complete transgene integration can be achieved by targeting 5′ and 3′ transgene sequences for PCR amplification. These target sequences should flank the complete GOI.
14. As the expected number of transgenic loci increases then the number of T2 progenies to be tested should be proportionately larger.
15. T2 progeny testing is a robust methodology for validating T1 zygosity determination by qPCR.
16. Soil mix in MK12 punnets can be used instead. In this case small volume of soil will restrict plant biomass production and ultimately seed yield. Typical wheat and barley seed yields are 4–10 g (when using 2.8 L 6 inch diameter pots).
17. The number of grains and spikelets per spike as well as spike length can be included in yield analysis.

References

1. Kohli A, Twyman RM, Abranches R et al (2003) Transgene integration, organization and interaction in plants. Plant Mol Biol 52:247–258
2. Svitashev SK, Somers DA (2001) Genomic interspersions determine the size and complexity of transgene loci in transgenic plants produced by microprojectile bombardment. Genome 44:691–697
3. Vain P, James A, Worland B et al (2002) Transgene behavior across two generations in a large random population of transgenic rice plants produced by particle bombardment. Theor Appl Genet 105:878–889
4. Forsbach A, Schubert D, Lechtenberg B et al (2003) A comprehensive characterization of single-copy T-DNA insertions into the Arabidopsis thaliana genome. Plant Mol Biol 52:161–176
5. Zadoks JC, Chang TT, Konzak CF (1974) A decimal code for the growth stages of cereals. Weed Res 14:415–421
6. Wilson AK, Latham JR, Steinbrecher RA (2006) Transformation-induced mutations in transgenic plants: analysis and biosafety implications. Biotechnol Genet Eng 23:209–234

Index

A

B

C

D

E

F

G

H

Delphine Fleury and Ryan Whitford (eds.), *Crop Breeding: Methods and Protocols*, Methods in Molecular Biology, vol. 1145, DOI 10.1007/978-1-4939-0446-4, © Springer Science+Business Media New York 2014

Printed by Publishers' Graphics LLC